Abdel Fattah A. Qaraman

Chemistry for Nursing

Abdel Fattah A. Qaraman

Chemistry for Nursing

An Applied Approach for Health Professions

Noor Publishing

Imprint

Cover image: www.ingimage.com

Publisher:
Noor Publishing
is a trademark of
Dodo Books Indian Ocean Ltd. and OmniScriptum S.R.L publishing group

120 High Road, East Finchley, London, N2 9ED, United Kingdom
Str. Armeneasca 28/1, office 1, Chisinau MD-2012, Republic of Moldova, Europe
Managing Directors: Ieva Konstantinova, Victoria Ursu
info@omniscriptum.com

Printed at: see last page
ISBN: 978-620-0-04514-0

Chemistry
for Nursing
Dr. Abdel-Fattah A. Qaraman
CH3CH2OH
NH4+
H2O
CO2
Cl

Chemistry for Nursing

Chemistry for Nursing

An Applied Approach for Health Professions

Dr. Abdel Fattah A. Qaraman

Associate Professor of Analytical and Inorganic Chemistry

Israa University, Palestine

2025

CHEMISTRY FOR NURSING

First edition. July 15, 2025.

Written by Abdel Fattah A. Qaraman.

To my

beloved wife, who has always been my support throughout

the journey of knowledge and work,

To my family, who gave me strength and unwavering support,

To my university, which nurtured my passion for learning,

And to my people, who always deserve the best.

I dedicate this humble work, hoping it contributes to the

advancement of science and service to humanity.

Dr. Abdel Fattah A. Qaraman

Understanding chemistry is essential to grasp the complex processes that sustain life and health

Table of Contents

Preface to the First Edition

Chemistry is often seen as a complex and abstract science, distant from nursing practice. Yet, it is fundamental to clinical decisions—from interpreting labs to preparing IV solutions and managing electrolytes.

Chemistry for Nursing was created to present chemical principles in a practical, clinically relevant way for nursing and allied health students. This edition bridges basic science and patient care, focusing on why chemistry matters at the bedside.

Chapters are clear and supported by clinical examples, case studies, step-by-step calculations, and visuals that connect theory to practice. Special emphasis is placed on safety and preparation, reflecting nurses' key role in therapeutic accuracy. This book aims to make chemistry accessible and relevant, helping students confidently apply concepts in real clinical situations.

Appendices include conversion tables, practice problems, safety tips, and resources to reinforce learning. We hope this text supports academic success and safe, evidence-based nursing care.

Feedback from educators and students is welcomed to improve future editions.

With appreciation,

Dr. Abdel Fattah A. Qaraman

About the book

Chemistry is often seen as an abstract or purely laboratory-based science. However, in healthcare, it plays a central and highly practical role in daily clinical practice. For students entering health professions such as nursing or pharmacy assistance, a solid foundation in chemistry is not merely helpful—it is essential.

In the human body, chemistry governs everything: from the oxygen we breathe and the nutrients we absorb, to the medications we administer and the electrical signals transmitted between nerve cells. Health professionals must understand these chemical processes to make informed clinical decisions, deliver safe and effective care, and communicate accurately with interdisciplinary teams.

This textbook, *Chemistry for Nursing*, is specially designed for first-year students in nursing and allied health programs. It presents chemistry in an accessible and engaging way, using real-world healthcare examples to demonstrate its relevance. Topics such as body temperature regulation, fluid and electrolyte balance, drug interactions, and metabolic pathways are explained in clear language with minimal reliance on advanced mathematics.

Key features of this book include:

- **Applied clinical focus** – Every chapter connects chemical concepts directly to medical and nursing practice.
- **Visual learning tools** – Diagrams, tables, and examples illustrate the link between theory and patient care.
- **Problem-solving skills** – Exercises help students master calculations involving dosages, concentrations, and pH.
- **Critical thinking development** – End-of-chapter questions encourage reflection, integration, and application of knowledge.

Whether you are preparing IV solutions, interpreting lab results, monitoring electrolytes, or learning the fundamentals of pharmacology, this book will provide the chemical understanding you need to perform your responsibilities confidently and safely.

By the end of this course, you won't just know the chemistry behind healthcare—you'll understand its value, its practical applications, and how to use it effectively in real clinical settings.

How to Study Chemistry for Nursing

Chemistry is a fundamental science for understanding how the human body functions, how medications work, and how healthcare interventions are developed and applied. For students in nursing and allied health programs, studying chemistry is not just about memorizing formulas—it's about applying critical thinking to real-life health situations.

This chapter will help you build a strong foundation for studying chemistry effectively. You will learn study strategies, how to approach problem-solving, and how to connect chemistry concepts to clinical settings.

✔ Why Chemistry Matters in Health Professions:

- Understand how drugs interact with the body at the molecular level.
- Learn how the body maintains balance through chemical reactions.
- Interpret lab results and understand medication mechanisms accurately.

✔ Study Strategies for Chemistry:

- Focus on understanding concepts rather than memorizing isolated facts.
- Use flashcards for common ions, symbols, and functional groups.
- Draw diagrams and concept maps to visualize reactions and processes.
- Practice solving problems regularly to build confidence.

✔ Problem-Solving Tips:

- Break down each problem into steps.
- Keep track of units in every calculation.
- Use dimensional analysis for conversion problems.
- Don't hesitate to go back and review basic math skills if needed.

✔ Tools and Resources:

- Periodic Table of Elements

- Calculator with scientific functions
- Study groups and peer discussions

By mastering these study techniques, you will be better prepared to succeed in this course and apply your knowledge in clinical practice.

Chapter 1

Chemistry in Human Health and Clinical Care

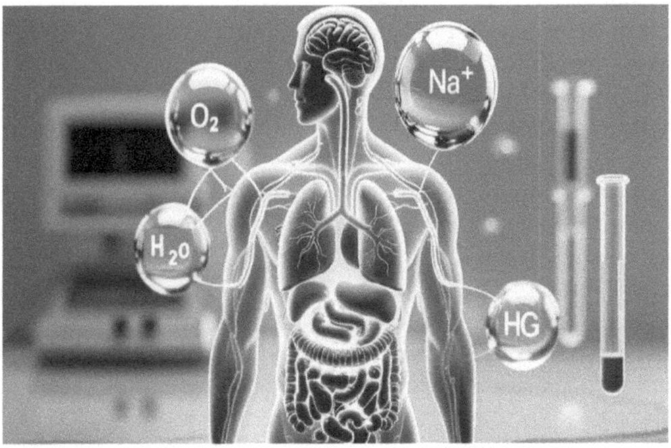

Chapter1 : Chemistry in Human Health and Clinical Care

Learning Objectives

By the end of this chapter, you should be able to:

1. Trace the historical development of chemistry in healthcare.
2. Identify major elements and ions essential to body function.
3. Explain the relevance of chemistry in nursing roles.
4. Outline the stages of drug development.
5. Apply chemical principles to patient care.

Chemistry is not merely a laboratory science—it is the silent foundation of modern healthcare. It explains how the body functions at the molecular level, encompassing essential processes such as breathing, digestion, fluid balance, and drug action. For nurses, a solid understanding of chemical principles is crucial for making safe clinical decisions, administering medications accurately, and monitoring patients effectively.

Whether preparing intravenous fluids, interpreting laboratory results, or managing electrolyte imbalances, chemistry underpins nearly every nursing intervention. Historically, it has transformed healthcare through milestones such as the discovery of antiseptics, antibiotics, and the synthesis of vaccines and complex pharmaceuticals. Today, it continues to drive innovation in areas like targeted drug delivery, personalized medicine, and biochemical diagnostics.

This chapter explores the essential relationship between chemistry and human health. It provides the scientific foundation necessary for nursing practice and demonstrates how chemical knowledge enhances clinical reasoning, supports patient safety, and improves healthcare outcomes.

The Historical Evolution of Chemistry in Health

Chemistry has influenced human health for thousands of years, both positively and negatively. Although formally recognized as a science in the 17th–18th centuries, its applications in healing date back to ancient times—particularly in traditional medicine, agriculture, and environmental health. Indigenous cultures made major contributions to early healthcare through their knowledge of plants, minerals, and animal-derived substances. Trial, error, and observation led to the use of thousands of natural remedies—many still used today.

One of the oldest medical documents, the *Ebers Papyrus* (circa 1550 BCE, Egypt), lists over 7,000 remedies in 800 formulas. Later, *De Materia Medica* by Dioscorides remained a core text for over 1,500 years.

During the Islamic Golden Age, Arab scientists such as Al-Razi and Ibn Sina advanced chemistry and medicine, introducing over 400 new drugs and developing techniques like distillation, crystallization, and filtration—many of which persist in modern pharmacy.

In Renaissance Europe, scholars like Paracelsus emphasized chemical treatments over superstition, laying foundations for pharmacology. The printing press accelerated the spread of scientific knowledge.

The 17th–18th centuries witnessed rapid scientific progress. Inventions like the microscope and thermometer sparked new disciplines—microbiology, immunology, and preventive medicine.

By the 19th century, active ingredients such as morphine and quinine were isolated, but high-purity drug production became feasible only in the mid-20th century through analytical chemistry.

From antiseptics like iodine to modern antibiotics, today's pharmacological tools reflect a rich chemical heritage. For nurses, appreciating this history supports safer practice and evidence-based care.

Figure 1.1: A scene from an ancient Egyptian papyrus showing a woman giving birth with the aid of attendants and protective deities

This image illustrates Egyptian childbirth scene from the Ebers Papyrus (ca. 1550 BCE), depicting women assisting in labor. This ancient medical manuscript not only described therapeutic substances but also recorded healing rituals. Many of these practices reflect early integrative approaches to medicine, blending natural remedies, observation, and cultural traditions—some of which inform today's complementary and holistic healthcare.

The Importance of Chemistry in Nursing and Healthcare

In modern healthcare, chemistry forms the foundation for understanding biological processes and pharmacological interactions in the human body, making it essential for all health professions—especially nursing.

Nurses regularly work with medications, IV fluids, and treatment protocols that require knowledge of chemical properties, drug compatibility, and side effect profiles. A strong background in chemistry equips nurses to:

1. Accurately calculate and administer drug dosages.
2. Understand drug–drug and drug–food interactions.
3. Handle hazardous or flammable substances safely in clinical environments.
4. Interpret laboratory test results and monitor patients' responses to therapy.

Biochemistry and pharmacology are core areas in nursing education. They empower nurses to make evidence-based decisions that directly affect patient outcomes. For example, administering insulin requires knowledge of its biochemical action and potential interactions with other treatments.

In short, chemistry is not merely theoretical—it is the practical language of life and the cornerstone of therapeutic interventions. Nurses, as frontline caregivers, must master this language to deliver safe and effective care.

What is Chemistry?

Chemistry is the science of matter—it studies the composition, structure, properties, and transformations of substances. In healthcare, chemistry helps explain how the body functions on a molecular level, how drugs exert their effects, and how nutrients and toxins are metabolized and eliminated.

Chemistry in Body Systems

Chemistry plays a fundamental role in both the structure and function of the human body. Every biological process is governed by chemical principles. The body is primarily composed of elements such as oxygen, carbon, hydrogen, nitrogen, calcium, and phosphorus, which form the essential molecules required for life.

Key ions—including sodium (Na^+), potassium (K^+), calcium (Ca^{2+}), and chloride (Cl^-)—act as electrolytes that support critical physiological processes such as nerve transmission, muscle contraction, fluid distribution, and maintaining electrical gradients.

Water, the most abundant compound in the body, serves as a universal solvent due to its polarity. It enables chemical reactions, transports nutrients, and helps regulate body temperature. Blood pH is tightly controlled between 7.35 and 7.45 by buffering systems such as the bicarbonate buffer, which maintain acid–base homeostasis.

Metabolic reactions, collectively known as metabolism, include **anabolic** processes that build complex molecules and **catabolic** processes that break them down to release energy. These reactions are catalyzed by **enzymes**, which accelerate reaction rates without being consumed.

The body relies on **macromolecules** such as:

- **Carbohydrates** for energy,
- **Proteins** for structure and enzyme function,
- **Lipids** for energy storage and membrane formation, and
- **Nucleic acids (DNA and RNA)** for storing genetic information.

Chemical signaling via **hormones and neurotransmitters** enables coordination between systems. For example, **insulin** regulates glucose metabolism, while **acetylcholine** is essential for nerve signaling.

Gases like **oxygen** and **carbon dioxide** are vital for cellular respiration. They are transported in the blood by proteins such as **hemoglobin**, and carbon dioxide also plays a key role in regulating blood pH.

Why This Matters in Nursing

A solid understanding of these chemical foundations enables healthcare professionals to interpret laboratory results, recognize conditions such as electrolyte imbalances, acid–base disturbances, and metabolic or respiratory dysfunctions, and apply this knowledge to safe and effective patient care.

These core concepts will be explored in greater depth in the chapters that follow.

Elements and Ions in the Human Body

The human body is composed of a range of chemical elements, each with specific roles essential to life and health. Among these, six elements—oxygen, carbon, hydrogen, nitrogen, calcium, and phosphorus—make up approximately 99% of total body mass. Oxygen, the most abundant element, is vital for cellular respiration and is primarily found in water and organic molecules. Carbon forms the structural framework of all organic compounds, including carbohydrates, proteins, lipids, and nucleic acids. Hydrogen, also abundant in water and biological molecules, contributes to acid-base balance and is involved in numerous metabolic processes. Nitrogen is a key component of amino acids and nucleotides, which are essential for protein synthesis and the formation of genetic material. Calcium is required not only for building bones and teeth but also for muscle contraction, blood clotting, and nerve transmission. Phosphorus, found mainly in bones and teeth, plays a critical role in energy transfer through ATP and is involved in the formation of DNA and RNA.

In addition to these major elements, the human body relies on several important ions—charged atoms or molecules—that support a variety of physiological functions. Sodium and potassium are crucial for maintaining electrochemical gradients across cell membranes, which enable nerve impulses and muscle contraction. Calcium ions are involved in neurotransmitter release, enzyme activation, and hemostasis. Chloride helps regulate osmotic balance and contributes to the production of hydrochloric acid in the stomach. Magnesium supports hundreds of enzymatic reactions and is essential for both muscular and neural regulation.

These elements and ions are tightly regulated within the body, and even slight imbalances can lead to significant disruptions in cellular activity, signaling pathways, or fluid homeostasis. Their essential roles highlight the importance of chemical principles in clinical practice, particularly in fields like nursing where ongoing monitoring and correction of such imbalances is part of everyday care.

Case Example

A patient presents with symptoms of muscle weakness and an irregular heartbeat. Upon reviewing the laboratory results, the healthcare provider identifies hypokalemia, a condition characterized by low potassium levels in the blood. Potassium (K^+) is a critical ion involved in maintaining the electrical excitability of nerve and muscle cells, including cardiac muscle. It plays a key role in generating action potentials and regulating heart rhythm. A deficiency in potassium disrupts this balance, leading to impaired muscle contraction and potentially life-threatening arrhythmias. By understanding the essential role of potassium in cellular function, the healthcare provider can promptly initiate appropriate interventions, such as potassium supplementation and monitoring, to restore normal levels and prevent further complications.

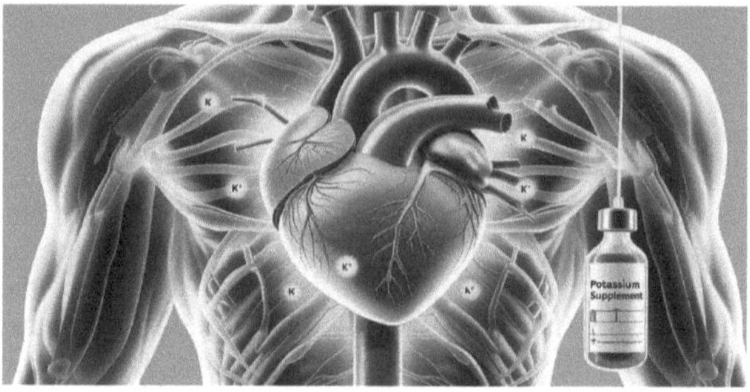

Figure 1.2 Potassium (K^+) plays a key role in regulating heart rhythm

Natural Sources of Medicines

Many modern medicines originate from natural sources, including plants, microorganisms, marine organisms, and minerals. Throughout history, humans have relied on natural products for healing, and today, a significant number of pharmaceuticals are still derived from or inspired by compounds found in nature. For example, the pain-relieving drug morphine is extracted from the opium poppy (*Papaver somniferum*), while aspirin was originally derived from salicin found in willow bark. Antibiotics such as penicillin come from fungi like *Penicillium* species, and streptomycin was discovered in soil bacteria (*Streptomyces*). The chemotherapy drug paclitaxel (Taxol®) was first isolated from the Pacific yew tree, and the antimalarial agent artemisinin comes from the sweet wormwood plant (*Artemisia annua*). Marine organisms like sponges, sea snails, and algae have also produced promising compounds for treating cancer, inflammation, and viral infections. Additionally, minerals such as magnesium sulfate (Epsom salt) and lithium salts are used to treat medical conditions ranging from constipation to bipolar disorder. These examples highlight the rich chemical diversity found in nature, which continues to serve as a valuable source of bioactive compounds and a foundation for drug discovery and development.

Drug Development and Chemistry

The process of drug development is deeply intertwined with chemistry at every stage—from initial molecular design to clinical application. Medicinal chemists play a central role in this process by designing, synthesizing, and optimizing chemical compounds to become safe and effective therapeutic agents. Their goal is to enhance a drug's beneficial effects while minimizing toxicity and unwanted side effects. This involves precise manipulation of chemical structures to improve the drug's potency, selectivity, stability, and solubility—qualities that determine whether a drug will be successful in real-world clinical settings.

A drug's effectiveness is governed by two key concepts: **pharmacokinetics** and **pharmacodynamics**. *Pharmacokinetics* (often summarized as ADME) refers to the body's effect on the drug—how it is **Absorbed**, **Distributed** through tissues, **Metabolized** (primarily in the liver), and ultimately **Excreted**. *Pharmacodynamics*, on the other hand, describes the drug's effect on the body—specifically how it binds to receptors or enzymes to produce a physiological response. Both processes are governed by fundamental chemical interactions at the molecular level.

For example, the chemical polarity and molecular size of a drug influence its ability to cross cell membranes, reach its target site, and undergo metabolism. Similarly, the shape and charge of a drug molecule determine its binding affinity to specific receptors, affecting both the strength and duration of its therapeutic action. Small changes in chemical structure can profoundly alter a drug's activity, side effect profile, or interaction with other medications.

Understanding these principles is not only important for pharmaceutical scientists but also essential for nurses. In clinical practice, nurses are responsible for administering medications, observing therapeutic responses, adjusting dosages, and educating patients about drug interactions and side effects. A working knowledge of drug chemistry empowers nurses to recognize when a medication might not be absorbed properly, when metabolism could be altered due to liver disease, or when multiple drugs may compete for the same metabolic pathway—leading to dangerous interactions.

Ultimately, chemistry ensures that medications are not only effective but also personalized—tailored to the unique physiological, genetic, and metabolic needs of each patient. By integrating chemical principles into patient care, nurses become critical partners in promoting medication safety, optimizing treatment outcomes, and advancing the science of pharmacology in everyday clinical practice.

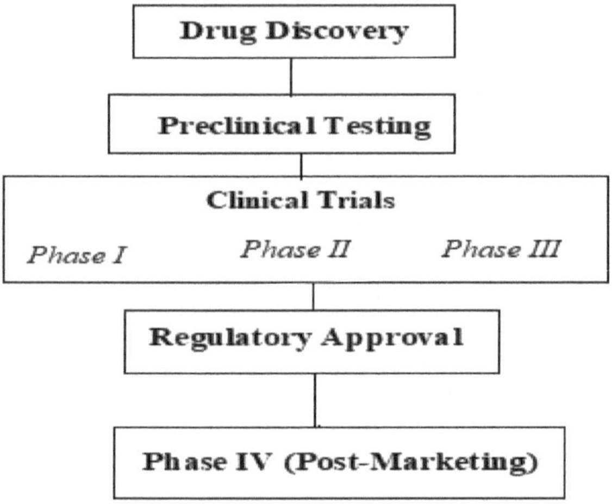

Figure 1.3 : Drug Development Process

1. **Drug Discovery:** Identification of active compounds from natural or synthetic sources.
   2. Preclinical Testing: Laboratory and animal studies to assess safety and biological activity.
   3. Clinical Trials:
   • *Phase I* – Safety testing in a small group of healthy volunteers.
   • *Phase II* – Assessment of effectiveness and side effects in patients.
   • *Phase III* – Large-scale trials to confirm efficacy and monitor adverse reactions.
   4. Regulatory Approval: Data reviewed by agencies (e.g., FDA, EMA) for safety and efficacy.
   5. Phase IV (Post-Marketing): Ongoing surveillance of long-term effects in the general population.

Table 1.1 highlights commonly used drugs, their origins, medical applications, and key chemical features that influence their biological activity.

Table 1.1 Examples of Common Drugs and Their Chemical Properties

Drug	Source	Use	Chemical Insight
Aspirin	Natural (willow bark)	Pain relief, anti-inflammatory	Ester group; inhibits COX enzymes, lowers prostaglandins
Paracetamol	Synthetic	Pain relief, fever reducer	Amide group; acts on hypothalamus, reduces fever
Atorvastatin	Synthetic (statin class)	Lowers cholesterol	Inhibits HMG-CoA reductase; lipophilic for liver targeting
Morphine	Natural (opium poppy)	Severe pain management	Alkaloid; binds mu-opioid receptors; crosses BBB
Penicillin	Natural (mold)	Bacterial infections	Beta-lactam ring; inhibits bacterial cell wall synthesis
Metformin	Synthetic (biguanide)	Type 2 diabetes	Biguanide structure; reduces liver glucose production, improves insulin sensitivity
Ibuprofen	Synthetic (propionic acid derivative)	Pain relief, anti-inflammatory, fever	Carboxylic acid group; inhibits COX-1/2, more selective than aspirin

Chemistry Link to Health

Translating Molecular Insights into Tomorrow's Clinical Solutions

Recent advancements in chemistry are deeply transforming the landscape of healthcare. Researchers are now using molecular-level insights to design **smart drug delivery systems**, **biocompatible materials**, and **personalized treatment strategies** based on an individual's genetic and metabolic profile. For example, Nanochemistry is enabling the delivery of chemotherapy drugs directly to cancer cells while sparing healthy tissues, greatly improving patient outcomes.

Another major development is the use of **chemical biomarkers** and **biosensors** for early diagnosis and monitoring of chronic diseases such as diabetes, cardiovascular disorders, and neurodegenerative conditions. These technologies offer rapid, point-of-care testing, allowing clinicians to intervene early and more effectively.

Looking to the future, chemistry will play a vital role in tackling global health challenges such as antibiotic resistance, aging-related diseases, and pandemic preparedness. By integrating artificial intelligence with chemical research, scientists aim to accelerate drug discovery and develop safer, more effective therapeutics.

Chapter summary

Chemistry is fundamental to human health, influencing traditional medicine, agriculture, and modern healthcare throughout history. Early humans used natural chemical substances from plants, animals, and minerals, with knowledge passed down through indigenous practices and ancient texts like the Ebers Papyrus. Advances by Arab physicians and European scientists during the Renaissance further developed pharmaceutical techniques and drug knowledge.

In modern healthcare, chemistry is crucial for understanding how the body functions at the molecular level, including drug interactions, metabolism, and physiological processes. Nurses and healthcare professionals rely on chemistry to safely administer medications, interpret lab results, and manage treatments effectively.

The human body's composition involves key elements (oxygen, carbon, hydrogen, nitrogen, calcium, phosphorus) and ions (sodium, potassium, calcium, chloride) essential for cellular functions, nerve impulses, muscle contraction, and acid-base balance.

Many medicines originate from natural sources such as plants (morphine, aspirin), fungi (penicillin), and marine organisms. Drug development combines chemistry and biology to create safe and effective treatments through a rigorous process of discovery, testing, and approval.

Overall, chemistry is the foundation of health sciences and indispensable in clinical care, especially nursing.

EXERCISES

Multiple Choice Questions:

1. Which ancient Egyptian medical text, dating to around 1550 BCE, contains descriptions of over 7,000 medicinal substances?

A) Hippocratic Corpus B) Ebers Papyrus

C) De Materia Medica D) Canon of Medicine

2. What is the main role of potassium (K^+) in the human body?

A) Bone formation B) Regulating nerve impulses and muscle contraction

C) Oxygen transport D) Digestive enzyme activation

3. Which of the following elements makes up the largest percentage by mass of the human body?

A) Nitrogen B) Calcium

C) Oxygen D) Phosphorus

4. Penicillin is derived from which natural source?

A) Willow bark B) Opium poppy

C) Mold (Penicillium notatum) D) Pacific yew tree

5. Which phase of clinical trials focuses on confirming effectiveness and monitoring adverse reactions in large populations?

A) Phase I B) Phase II C) Phase III D) Phase IV

Short Answer Questions:

1. Explain the significance of the bicarbonate buffer system in the human body.
2. Describe the role of enzymes in chemical reactions within the body.
3. List and briefly describe the main stages involved in the drug development process.
4. How did Arab physicians historically contribute to the advancement of chemistry in medicine?
5. Why is a strong understanding of chemistry essential for nursing professionals?

Case-Based Problem: A hospitalized patient presents with muscle weakness and an irregular heartbeat. Laboratory tests confirm hypokalemia (low potassium levels). • Explain how potassium deficiency affects muscle and heart function. • Suggest possible interventions based on chemical knowledge.

True or False:

1. Morphine is a naturally occurring compound derived from the opium poppy.
2. The human body's acid-base balance is maintained by buffer systems like bicarbonate.
3. Chemistry only became relevant to human health after the invention of the microscope.
4. Magnesium ions are important for enzymatic reactions in the body.
5. Penicillin contains a beta-lactam ring that disrupts bacterial cell walls.

Fill in the Blanks:

1. Chemistry is the science that studies the ________, ________, ________, and ________ of matter.
2. The most abundant elements in the human body include ________, ________, ________, ________, ________, and ________.
3. Potassium ions (K^+) play a key role in regulating ________ and ________ contraction, including the heart muscle.

4. One of the oldest known medical texts containing thousands of medicinal substances is the ________ Papyrus.
5. The drug morphine is extracted from the ________ poppy plant.
6. In the drug development process, the phase where safety is tested in a small group of healthy volunteers is called Phase ________.
7. Iodine (I_2) was introduced in the 19th century as a disinfectant for treating ________ and preventing infections.
8. ________ is the primary solvent in the human body, important for nutrient transport and temperature regulation.
9. Enzymes speed up chemical reactions in the body without being consumed in the process, making them ________.
10. The major ions involved in fluid balance in the body include ________ and ________.

Bonus Application Question: A nurse needs to administer 40 mEq of potassium chloride IV. The available solution contains 20 mEq/10 mL. How many milliliters should be given?

Chapter 2

Measurements and Biomedical Applications

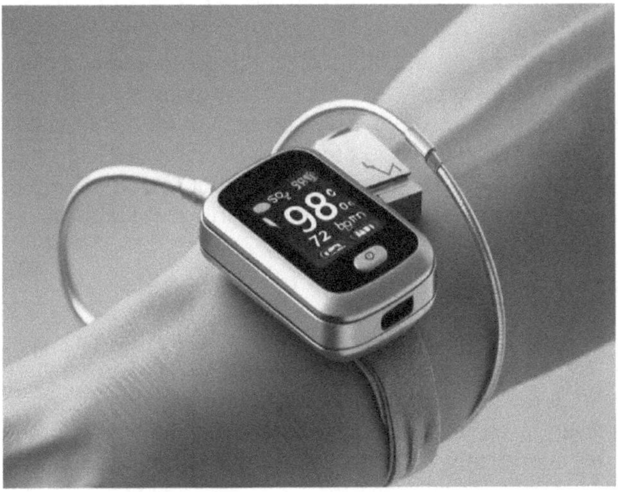

Chapter 2 :Measurements and Biomedical Applications

Learning Objectives

By the end of this chapter, you should be able to:

1. Understand and convert between metric and customary units.
2. Apply temperature and pressure concepts to clinical scenarios.
3. Analyze pH balance and interpret basic acid–base disorders.
4. Calculate caloric values and interpret energy needs.

In the field of nursing and biomedical sciences, precise measurement is not merely a technical skill—it is a cornerstone of safe, effective, and evidence-based patient care. From administering medications and monitoring vital signs to preparing IV solutions and interpreting laboratory results, nurses rely on an accurate understanding of units, conversions, and measurement principles to make informed clinical decisions.

This chapter provides a comprehensive overview of chemical measurements and their direct relevance to health care settings. It begins by reviewing the **metric and U.S. customary systems**, highlighting the essential units of length, mass, volume, temperature, and pressure used in clinical environments. Students will also learn how to confidently convert between units and apply these conversions in practical scenarios.

Next, the chapter introduces critical concepts such as **significant figures**, **scientific notation**, **accuracy**, **precision**, and **percent error**—all of which are essential in laboratory diagnostics and data interpretation. These principles serve as a foundation for evaluating the reliability of clinical data and preventing errors in medication dosing or diagnostic assessments.

Special focus is given to the biomedical importance of physical properties such as **density**, **temperature**, and **pressure**, with clinical examples that illustrate their roles in hydration status, blood circulation, and respiratory function. Additionally, the chapter explores the **pH scale and acid–base balance**, a key aspect of homeostasis with major implications for metabolic and respiratory conditions.

Finally, the chapter concludes with a discussion on **energy and nutrition**, introducing how the body utilizes macronutrients and the biochemical basis of calorie metabolism—information vital for nursing care in both acute and chronic conditions.

By mastering these measurement concepts, nursing students will be better equipped to deliver safe, accurate, and scientifically grounded care across diverse clinical situations.

Units of Measurement and Conversions

In nursing and biomedical sciences, the ability to understand and accurately use different systems of measurement is essential. From calculating medication dosages to interpreting laboratory values and preparing intravenous (IV) fluids, precise measurement is a critical skill that directly affects patient safety. This section introduces both the **metric system**, which is the international standard in scientific and medical contexts, and the **U.S. customary system**, which is still used in various aspects of clinical and home care settings.

The Metric System

The **metric system** is a decimal-based system of measurement that is widely used in healthcare due to its simplicity and consistency. It uses base units that can be scaled up or down using prefixes that represent powers of ten.

Table 2.1 Common Metric Units:

Quantity	Unit	Symbol
Length	meter	m
Mass (Weight)	gram	g
Volume	liter	L
Temperature	degrees Celsius	°C

Table 2.2 Metric Prefixes:

Prefix	Symbol	Value
kilo	k	1,000 units
centi	c	0.01 units
milli	m	0.001 units
micro	μ	0.000001 units

Examples:

- 1 kilogram (kg) = 1,000 grams (g)
- 1 milliliter (mL) = 0.001 liter (L)

These units are standard in medication preparation, lab results, and physiological measurements such as blood pressure, body temperature, and fluid volumes.

U.S. Customary (Household) Units

While the metric system is predominant in professional healthcare, **U.S. customary units** (also referred to as household units) are still frequently encountered in community settings, patient instructions, over-the-counter (OTC) medications, and dietary planning.

Table 2.3 Common U.S. Units:

Quantity	Unit	Abbreviation
Length	inch	in
Weight	pound	lb
Volume	teaspoon	tsp
	tablespoon	Tbsp
	ounce	oz
	cup	cup
Temperature	degrees Fahrenheit	°F

Common Conversions Between Metric and U.S. Units

Healthcare professionals frequently convert between metric and U.S. customary units when caring for patients from different regions or interpreting various medical resources. Below are some essential conversion values:

Length:

- 1 inch = 2.54 centimeters (cm)
- 1 foot = 12 inches = 30.48 cm

Mass/Weight:

- 1 pound (lb) = 16 ounces = 454 grams (g)
- 1 kilogram (kg) = 2.2 pounds (lb)

Volume:

- 1 teaspoon (tsp) = 5 milliliters (mL)
- 1 tablespoon (Tbsp) = 15 mL
- 1 fluid ounce (oz) = 30 mL
- 1 cup = 8 oz = 240 mL
- 1 pint = 16 oz = 480 mL
- 1 quart = 32 oz = 960 mL
- 1 gallon = 128 oz = 3840 mL

Temperature:

- F^0 to C^0 : $C^0 = (F0-32) \div 1.8$
- C^0 to F^0 : $F^0 = (C^0 \times 1.8)+32$

These conversions are especially useful in international healthcare environments and in patient education, where information must be communicated clearly across systems.

Clinical Significance of Measurement Systems

Nurses must be fluent in both measurement systems to:

- Accurately calculate and administer medications.
- Convert patient-reported intake/output using household terms.
- Provide dietary and fluid recommendations.
- Ensure safe communication across different healthcare settings and cultural contexts.

For instance, a patient may be instructed at home to drink "8 cups of water" per day, while a nurse needs to document this as **1,920 mL** in a hospital chart.

Best Practices for Nurses

- **Always double-check conversions**, especially for vulnerable populations such as pediatric or geriatric patients.
- Avoid using outdated or ambiguous abbreviations (e.g., use "mL" instead of "cc").
- Educate patients using visual aids (e.g., dosing cups or oral syringes) to improve safety and understanding.
- Use **dimensional analysis (factor-label method)** when performing conversions to reduce calculation errors.

Example:
To convert 0.5 grams (g) to milligrams (mg):

$0.5g \times (1000\ mg/1\ g) = 500\ mg$

This foundational understanding of units and conversions prepares nursing students to perform with confidence in both clinical and academic environments. Mastery of measurement systems not only supports accurate documentation and medication safety but also enhances communication with interdisciplinary teams and patients.

Key Takeaways

- The **metric system** is the standard for medical measurements; it is decimal-based and includes units like **meter (m), gram (g), liter (L), and degrees Celsius (°C).**
- **U.S. customary units** are still encountered in clinical practice, especially in patient education, OTC medications, and nutrition.
- Nurses must be proficient in converting between units using standard conversion factors.
- **Common conversions** include:
    - 1 tsp = 5 mL
    - 1 Tbsp = 15 mL
    - 1 oz = 30 mL
    - 1 kg = 2.2 lb
    - 1 inch = 2.54 cm

• Always ensure **clarity**, **accuracy**, and **patient safety** when documenting or communicating measurements.

Practice Problems: Convert the following measurements:

1. 2 kilograms = ______ grams
2. 3 teaspoons = ______ milliliters
3. 1.5 liters = ______ milliliters
4. 98.6°F = ______ °C
5. 180 pounds = ______ kilograms

Case Study: Pediatric Medication Dosage

A mother reports that she has been giving her child "2 teaspoons of cough syrup every 6 hours" as instructed on the bottle. The nurse must record this dosage in milliliters and verify if the dose is within the safe range based on the child's weight.

- **Step 1:** Convert the volume
  2 tsp × 5 mL/tsp = **10 mL**
- **Step 2:** Verify dosage safety
  If the safe dose is 1 mg/kg and the child weighs 15 kg, the maximum allowed is **15 mg per dose**.

If the syrup contains 1 mg/mL, the child is receiving **10 mg**, which is **within the safe range**.

Nursing Insight:
Conversions are not just mathematical — they directly affect whether a dose is therapeutic or toxic. Understanding units empowers nurses to protect patient safety.

Significant Figures and Scientific Notation

In healthcare and laboratory settings, numerical accuracy matters. Understanding **significant figures** ensures that measurements reflect the precision of the tools used. The number of significant figures in a value tells us how exact the measurement is.

Table 2.3 Rules of Significant Figures

Rule	Example	Explanation
1. All non-zero digits are significant	123	Has 3 significant figures
2. Zeros between non-zero digits are significant	1003	Has 4 significant figures
3. Leading zeros (before the first non-zero digit) are not significant	0.0025	Has 2 significant figures (zeros are placeholders)
4. Trailing zeros in a decimal number are significant	45.00	Has 4 significant figures (shows measurement precision)
5. Trailing zeros in a whole number without a decimal are not significant	1500	Has 2 significant figures
5b. Trailing zeros in a whole number **with a decimal** are significant	1500.	Has 4 significant figures (decimal point makes zeros significant)
6. Exact numbers have infinite significant figures	1 dozen = 12	Infinite precision (not measured, but defined or counted)

For example:

- 25.0 g has **three significant figures** (the trailing zero is significant).
- 0.0450 mg has **three significant figures** (leading zeros are not significant, but the trailing zero is).

Significant figures are important in dosage calculations, lab reporting, and conversions. Reporting too many or too few digits can mislead or endanger patient care.

Question: How many significant figures are in the measurement 1. 0.0450 mg 2. 7.001 L 3. 0.0008 g and why?

Scientific Notation
Scientific notation is a method used to express **very large** or **very small numbers** in a compact and readable format. It is especially useful in healthcare and laboratory science, where precision and clarity are essential.

Examples:
• 0.000003 g = $\mathbf{3 \times 10^{-6}}$ **g**
• 2,500,000 mL = $\mathbf{2.5 \times 10^{6}}$ **mL**

Scientific notation simplifies numerical expressions and improves clarity when recording **lab results**, **drug concentrations**, and **medical data**.

Question:

1: Convert to scientific notation

a) 0.00089 g = ?
b) 7,200,000 mL = ?

2: Convert from scientific notation to standard form

a) 3.2×10^{5} mg = ?
b) 6.5×10^{-3} L = ?

Accuracy and Precision in Clinical Measurement

In the context of clinical practice and biomedical analysis, the concepts of accuracy and precision are central to ensuring data validity, therapeutic safety, and diagnostic reliability. Despite being closely related, they refer to distinct aspects of measurement performance.

Accuracy is defined as the degree to which a measured value approximates the **true or accepted reference value**. High accuracy minimizes systematic error and enhances the clinical relevance of the measurement.

Precision, on the other hand, refers to the **reproducibility or consistency** of measurements when repeated under unchanged conditions. A measurement can be precise (i.e., tightly grouped results) without necessarily being accurate.

Illustrative Example:
Consider a thermometer:

- If it consistently records **98.6°F** when the actual body temperature is **98.6°F**, the measurement is both **accurate and precise**.
- If it yields values like **98.5°F**, **98.6°F**, and **98.7°F**, it is **precise** due to consistency, even if a systematic error exists.

In clinical and laboratory environments, achieving both accuracy and precision requires:

- Well-calibrated instruments
- Standardized protocols
- Technically competent personnel

These principles underpin reliable activities such as:
- Interpreting laboratory diagnostics
- Calculating and administering precise medication dosages
- Monitoring vital signs and physiological responses
- Conducting evidence-based clinical assessments

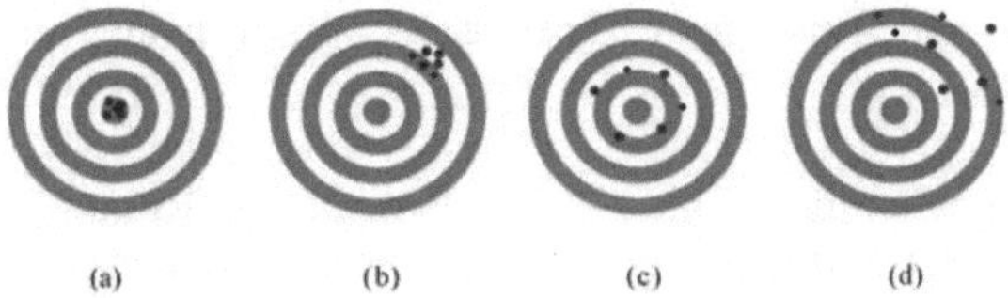

Figure 2.1 Accuracy vs. precision: (a) high accuracy, high precision; (b) low accuracy, high precision; (c) high accuracy, low precision; (d) low accuracy, low precision.

Percent Error

Percent error is a method used to evaluate the **accuracy** of a measured value by comparing it to a known or accepted standard. It is especially useful in laboratory settings to assess performance and identify possible sources of error.

$$percent\ error = \frac{\text{Measured Value} - \text{True Value}}{\text{True Value}} \times 100$$

Percent Error and Its Clinical Significance

In healthcare and biomedical science, **percent error** is used to evaluate the **accuracy** of experimental or clinical measurements by comparing them to a known or accepted standard.

A **low percent error** (typically $<5\%$) indicates a **high degree of accuracy** and suggests that the measurement is clinically acceptable. A **high percent error** may point to:

- Instrument calibration errors
- Improper technique
- Contaminated or degraded samples
- Mislabeling or transcription mistakes

Clinical significance:

Even small errors can lead to incorrect diagnoses or inappropriate treatments. For example, a miscalculated drug concentration could result in an underdose or overdose, especially in pediatric or critical care settings.

Therefore, percent error is not just a mathematical value—it is a **quality control tool** that supports **safe and effective clinical decision-making**.

Example: Calculating Percent Error
A laboratory technician measures the density of a biological fluid and obtains a value of **1.03 g/mL**. The **accepted (true) value** for this fluid is known to be **1.00 g/mL**.

To assess the accuracy of the measurement, the percent error is calculated using the formula:

$$percent\ error = \frac{1.03 - 1.00}{1.00} \times 100 = 3\%$$

I Interpretation: A 3% error indicates that the measurement is reasonably accurate, with only a small deviation from the true value.

Clinical Case Study: Percent Error in IV Medication

A nurse prepares to administer an intravenous medication to a neonate. The drug must be diluted to a final concentration of **0.80 mg/mL**. After preparing the solution, the pharmacy technician measures the concentration and finds it to be **0.92 mg/mL**.

Question:

1. Calculate the **percent error** between the measured and the intended concentration.
2. Based on the result, **is this deviation clinically acceptable**? Justify your answer.

Density and its Biomedical Significance

What is Density?

Density is a fundamental physical property that expresses the **mass of a substance per unit volume**. It is defined by the formula:

$$Density = \frac{\text{Mass}}{\text{Volume}}$$

- **SI unit**: kilograms per cubic meter (kg/m^3)
- **Common unit in healthcare**: grams per milliliter (g/mL) or grams per cubic centimeter (g/cm^3) *(Note: 1 g/mL = 1 g/cm³)*

Understanding density allows healthcare professionals to compare different substances and assess concentration, purity, or pathological changes in biological fluids.

Measuring Density in the Clinical Context

In a laboratory or clinical setting, density is often measured by determining the mass of a known volume of fluid, using tools like:

- **Graduated cylinders or pipettes** (for accurate volume measurement)
- **Electronic balances** (for precise mass determination)

For example:
If a nurse measures 50.0 mL of urine and its mass is 51.5 grams, the density is:

$$Density = \frac{51.5\ \text{g}}{50.0\ \text{ml}} = 1.03\ g/ml$$

Biomedical Significance of Density

Density has practical and diagnostic value in a variety of **healthcare and biomedical settings**, including:

1. Urine Specific Gravity

- A common **urinalysis test** evaluates urine density to assess **hydration status** or detect potential disorders such as **diabetes mellitus, kidney dysfunction**, or **urinary tract infections**.
- Normal range: **1.005 to 1.030**
  Higher values indicate **concentrated urine** (e.g., dehydration), while lower values may indicate **diluted urine** (e.g., overhydration or renal disease).

2. Plasma and Blood Density

- Plasma has a typical density of about **1.025 g/mL**.
- Changes in **blood density** may reflect variations in **lipid concentration, protein levels**, or **hematocrit**, affecting diagnosis of anemia or hyperlipidemia.

3. Drug Formulation and Dosage

- **Liquid medications** must be prepared based on density to ensure accurate **concentration per volume** (e.g., mg/mL), especially in IV solutions and pediatric dosing.

4. Body Composition Analysis

- **Densitometry**, such as hydrostatic weighing, is used to assess **body fat percentage** by comparing total body density with reference standards.

Quick Facts

- Water has a density of **1.00 g/mL** at 4°C — used as a baseline in many comparisons.
- Substances **denser than water** (e.g., bone, metals) sink, while **less dense** substances (e.g., fat, oil) float.
- The density of biological fluids is **temperature-dependent**, and accurate measurements should consider this factor.

Clinical Case Example

Case: A dehydrated patient presents with dark yellow urine. The nurse measures the urine's mass (52.0 g) and volume (48.0 mL).

$$Density = \frac{52.0\ \text{g}}{48.0\ \text{ml}} = 1.08\ g/ml$$

Interpretation: This abnormally high density confirms **urine concentration**, aligning with signs of dehydration. The nurse should initiate fluid replacement therapy and monitor urine output.

Clinical Question: IV Medication Preparation

A pediatric nurse is preparing an intravenous (IV) medication. The final volume is **25.0 mL**, and the measured mass of the solution is **26.5 grams**.

1. Calculate the density of the prepared solution.
2. Why is it important to know the density when calculating drug dosage for children?

Temperature Measurement

Temperature is a fundamental physical property that reflects the degree of heat present in a substance or environment. In healthcare, precise measurement of temperature is crucial for assessing a patient's physiological status, diagnosing infections, monitoring metabolic function, and regulating drug storage conditions.

What is Temperature?

Temperature quantifies the **average kinetic energy of particles** in matter. As particles move faster, temperature increases; as they slow down, temperature decreases. Unlike heat, which is energy transferred, temperature is a measure of that energy's intensity.

Common Temperature Scales

In clinical and scientific settings, temperature is typically measured using one of three main scales: Celsius (°C), Fahrenheit (°F), and Kelvin (K). Each scale has its specific applications depending on the context and region.

Figure 2.2 Water freezing and boiling poin

Table 2.4 Temperature scales

Scale	Freezing Point of Water	Boiling Point of Water	Unit Symbol	Usage in Healthcare
Celsius (°C)	0°C	100°C	°C	Standard in most countries and labs
Fahrenheit (°F)	32°F	212°F	°F	Predominantly in the United States
Kelvin (K)	273.15 K	373.15 K	K	Used in scientific research, rarely in clinical settings

Conversion Between Celsius and Fahrenheit

Converting between °C and °F is common in nursing practice, especially when reviewing medical literature or communicating with patients internationally.

- **From Fahrenheit to Celsius:**

$$°C = \frac{5}{9}(°F - 32)$$

- **From Celsius to Fahrenheit:**

$$°F = \frac{9}{5}(°C) + 32$$

Example: If a patient's temperature is 102.2 °F, what is it in Celsius?

Solution: Use the formula:

$$°C = \frac{5}{9}(°102.2\ °F\ - 32) = 39.0°C$$

Interpretation: A body temperature of 39.0 °C indicates **high fever**, which may suggest an **infection, inflammation**, or other underlying medical condition and requires prompt clinical evaluation.

Question: Convert the following temperatures:

1. Convert 98.6°F to Celsius.
2. Convert 37°C to Fahrenheit.

Methods of Temperature Measurement

In clinical settings, body temperature can be measured using several techniques, each with unique advantages and limitations:

1. **Oral Temperature**:
   Measured under the tongue. It is convenient and widely used but may be influenced by recent food or drink intake or breathing through the mouth.

2. **Axillary Temperature**:
   Taken under the armpit. It is less accurate but non-invasive and useful for infants or uncooperative patients.
3. **Tympanic (Ear) Temperature**:
   Measured with an infrared device aimed at the eardrum. It is fast and reasonably accurate, though affected by earwax or incorrect placement.
4. **Rectal Temperature**:
   Considered the most accurate method for core body temperature. Used especially in infants or critically ill patients.
5. **Temporal Artery Temperature**:
   Taken by scanning the forehead with an infrared thermometer. It is non-invasive, quick, and generally reliable.

Biomedical Importance of Accurate Temperature Measurement

- **Fever Detection**: Temperatures above 38 °C (100.4 °F) may signal infection or inflammation.
- **Hypothermia Monitoring**: Readings below 35 °C (95 °F) may indicate exposure or metabolic dysfunction.
- **Medication Dosage Adjustments**: Some medications require dosage changes based on body temperature.
- **Equipment Calibration**: Monitoring temperatures in lab equipment (e.g., vaccine refrigerators) ensures drug safety and efficacy.

Clinical Tips for Nurses

- Choose the measurement method carefully and consider its effect on accuracy.
- Be mindful of factors such as exercise, room temperature, or recent ingestion that can skew readings.
- Follow proper infection control protocols during measurement.
- Always document temperature, measurement method, site, and time.

Clinical Case Scenarios

- **Case 1: Fever Evaluation**
  A patient's oral temperature reads 101.3°F. Calculate the temperature in Celsius and explain whether this indicates a fever. What nursing interventions should be considered?
- **Case 2: Hypothermia Concern**
  An elderly patient in the emergency room has a rectal temperature of 34.5°C. What are the implications, and what immediate care steps should the nurse take?
- **Case 3: Vaccine Storage**
  The vaccine refrigerator is monitored at 5°C. Why is maintaining this temperature critical for vaccine efficacy?

Case 1: $°\mathrm{C} = \frac{5}{9}(°101.3\ °F\ - 32) = 38.5°C$

- This indicates a fever (normal is around 37°C).
- Nursing interventions: Monitor vitals regularly, provide antipyretics as ordered, ensure hydration, assess for infection source.

Case 2:

- Hypothermia (below 35°C or 95°F).
- Immediate care: Warm the patient gradually, monitor vital signs closely, prevent further heat loss, and assess for underlying causes.

Case 3: Maintaining 2°C to 8°C (36°F to 46°F) is essential to preserve vaccine potency and prevent spoilage.

Table 2.5 Temperature Conversion

°C (Celsius)	°F (Fahrenheit)	Comment
0	32	Freezing point of water
25	77	Typical room temperature
37	98.6	Normal human body temperature
38	100.4	Fever threshold
40	104	High fever
100	212	Boiling point of water

Pressure in the Human Body

What is Pressure?

Pressure is defined as the **force applied per unit area**. It is expressed mathematically as:

$$Pressure = \frac{\text{Force}}{\text{Area}}$$

- **SI unit:** Pascal (Pa), where

$$1Pa = \frac{1\ \text{N}}{\text{m}^2}$$

- Other units common in medicine: millimeters of mercury (mmHg), centimeters of water (cmH_2O)

Pressure plays a critical role in various physiological processes, from blood circulation to respiratory function.

Types of Pressure Relevant to Human Physiology

1. **Blood Pressure**
   The force exerted by circulating blood on the walls of blood vessels. It is a vital sign reflecting cardiovascular health.

   - Measured in mmHg
   - Expressed as systolic/diastolic pressure (e.g., 120/80mmHg)
   - Systolic: pressure during heart contraction
   - Diastolic: pressure during heart relaxation

2. **Intraocular Pressure (IOP)**
   Pressure inside the eye, crucial for maintaining eye shape and proper vision.

   - Normal range: 10–21 mmHg
   - Elevated IOP is a risk factor for glaucoma.

3. **Intracranial Pressure (ICP)**
   Pressure within the skull, affecting brain function.

- Normal ICP: 7–15 mmHg in adults
- Elevated ICP can indicate brain injury or swelling.

4. **Respiratory Pressure**

Breathing is driven by pressure differences between the atmosphere, lungs, and pleural cavity. The key types of respiratory pressure include:

- **Atmospheric Pressure:** The pressure exerted by the air surrounding the body, approximately **760 mmHg** at sea level.
- **Intrapulmonary Pressure:** The pressure within the alveoli of the lungs. It fluctuates with breathing: slightly below atmospheric pressure during inhalation (~759 mmHg), slightly above it during exhalation (~761 mmHg), and equal to atmospheric pressure (~760 mmHg) at the end of each breath when airflow stops.
- **Intrapleural Pressure:** The pressure within the pleural cavity, which remains negative relative to atmospheric pressure (about **-4 mmHg**). This negative pressure is essential for keeping the lungs inflated and preventing their collapse.

Measurement of Pressure in Clinical Settings

- **Sphygmomanometer:** Measures arterial blood pressure using a cuff and mercury or aneroid gauge.
- **Tonometer:** Measures intraocular pressure.
- **Intracranial pressure monitors:** Invasive devices used in critical care for patients with brain injuries.
- **Manometers and pressure transducers:** Used for respiratory pressures.

Biomedical Significance

- Maintaining appropriate pressures is essential for **homeostasis** and **organ function**.
- Abnormal blood pressure can lead to **hypertension** or **hypotension**, increasing risks for heart attack, stroke, or shock.

- Elevated IOP can cause **optic nerve damage**, leading to vision loss.
- Abnormal ICP requires immediate intervention to prevent brain damage.
- Understanding respiratory pressures helps in managing **ventilator settings** and treating lung diseases.

Clinical Case Studies

Case 1: Hypertension Alert

A 55-year-old male presents with a blood pressure reading of 168/98 mmHg. This indicates stage 2 hypertension, which significantly increases the risk of cardiovascular events such as stroke and heart disease.
Answer: The patient has stage 2 hypertension. Nursing interventions include lifestyle counseling (diet, exercise, sodium reduction), medication review, and regular blood pressure monitoring.

Case 2: Elevated Intracranial Pressure

A 32-year-old female with a history of head trauma arrives with symptoms including vomiting, confusion, and unequal pupils. Her ICP is recorded at 22 mmHg, which is above the normal range (7–15 mmHg).

Answer: The patient has elevated intracranial pressure. Immediate intervention is required to prevent brain herniation, including raising the head of the bed, oxygen support, and preparing for neurosurgical evaluation.

Case 3: Glaucoma Screening
An elderly female undergoing a routine eye examination has an intraocular pressure (IOP) reading of 24 mmHg. The normal range for IOP is 10–21 mmHg.
Answer: The patient shows signs of elevated IOP, suggesting possible early glaucoma. Referral to an ophthalmologist is necessary, and patient education on the importance of follow-up is recommended.

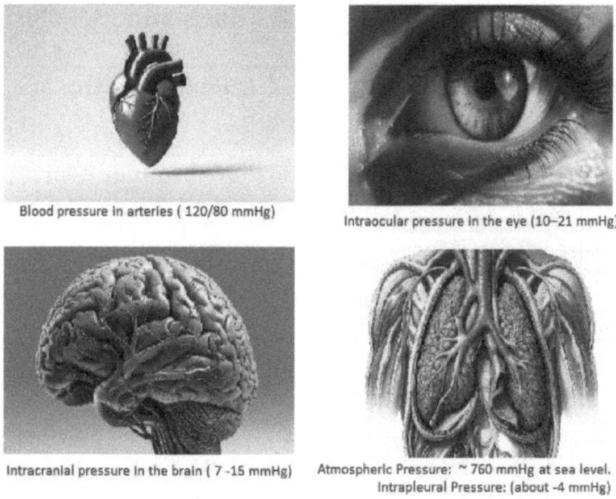

Figure 2.3 Pressures in the Human Body

pH and Acid–Base Balance

What Is pH?

pH is a measure of the **hydrogen ion concentration $[H^+]$** in a solution. It indicates how **acidic or basic (alkaline)** a substance is:

$$\textbf{pH} = \log\ [\boldsymbol{H}^+]$$

- **pH Scale:** Ranges from **0 to 14**
  - pH < 7 → **Acidic**
  - pH = 7 → **Neutral**
  - pH > 7 → **Basic (alkaline)**

In clinical practice, pH is vital for assessing the body's internal environment, especially blood and other bodily fluids.

Normal Blood pH

- **Normal arterial blood pH:** 7.35–7.45
- Values outside this range can disrupt cellular function and enzyme activity.

Table 2.6 Clinical Significance of Acidosis and Alkalosis

Condition	pH Range	Clinical Significance
Acidosis	< 7.35	Can depress CNS, impair heart function
Alkalosis	> 7.45	Can cause muscle spasms, confusion, arrhythmias

Example 1: Calculating pH from $[H^+]$

If the hydrogen ion concentration is:

$[H^+] = 1 \times 10^{-3}$ **mol/L**

Then: **pH = $-\log(1 \times 10^{-3}) = 3$**

→ The solution is **acidic** because pH < 7.

Example 2: Calculating $[H^+]$ from pH

If the pH of a solution is: **pH = 8**

Then the hydrogen ion concentration is: $[H^+] = 10^{-8}$ **mol/L**

→ The solution is **basic (alkaline)** because pH > 7.

Check Your Self:

1. Calculate the pH of a solution with $[H^+] = 1 \times 10^{-4}$ mol/L.
2. A patient's ABG results show a pH of 7.48. Is this considered alkalosis? Explain.

Energy and Nutrition Basics

What Is Energy?

In biological systems, **energy** is the capacity to do work — including muscle contraction, nerve signaling, and maintaining body temperature. The human body derives energy from **nutrients** through chemical reactions that occur primarily in cells.

- **Unit of energy:**
  - In nutrition: **kilocalories (kcal)**
  - In chemistry: **joules (J)**
    (1 kcal = 4.184 kJ)

Macronutrients as Energy Sources

The human body relies on three major macronutrients—**carbohydrates**, **proteins**, and **fats**—to meet its energy demands and maintain essential physiological functions. Each macronutrient provides a specific amount of energy per gram and serves unique biological roles. The table below summarizes their energy yield and functional significance in clinical nutrition.

Table 2.7 Macronutrients and their Role in the Body

Macronutrient	**Energy Yield (kcal/g)**	**Role in the Body**
Carbohydrates	4 kcal/g	Quick energy; brain's preferred fuel
Proteins	4 kcal/g	Building tissues; used as energy if needed
Fats	9 kcal/g	Long-term energy storage; insulation

Note : Alcohol also provides energy (7 kcal/g) but is not a nutrient and has no essential biological role.
How the Body Uses Nutrients for Energy

- Nutrients are broken down via **metabolism**, a series of chemical reactions.
- In **cellular respiration**, glucose is converted into **ATP (adenosine triphosphate)**, the body's energy currency.
- ATP powers all vital cellular processes.

Basic equation for glucose metabolism:

When the body breaks down **glucose ($C_6H_{12}O_6$)** in the presence of **oxygen (O_2)**, it produces **energy (in the form of ATP)**, along with **carbon dioxide (CO_2)** and **water (H_2O)** as byproducts. The overall chemical equation is:

$$C_6H_{12}O_6 + 6O_2 \rightarrow 6CO_2 + 6H_2O + ATP$$

Nutrition in Clinical Practice

Nutrition is a foundational aspect of patient care that directly affects healing, recovery, and long-term health outcomes. For nurses, understanding nutritional principles is not only essential for identifying deficiencies or excesses, but also for guiding clinical decisions related to energy needs, therapeutic diets, and specialized feeding methods. The following points highlight key areas where nutritional knowledge plays a critical role in nursing practice.

- **Monitoring calorie needs** in undernourished or critically ill patients.
- **Managing chronic conditions** like diabetes, obesity, and cardiovascular disease.
- **Promoting wound healing**, which requires adequate protein and energy intake.
- **Assessing enteral/parenteral feeding** and interpreting nutrition labels.

Understanding key terms related to energy and nutrition helps nurses assess patient needs, plan appropriate diets, and monitor metabolic health. The following table summarizes essential terms that support clinical decision-making in nutritional care.

Table 2.8 – Energetic Key Terms for Nurses

Term	Definition
Basal Metabolic Rate (BMR)	Energy needed at rest to maintain basic functions
Calorie	Unit of energy (technically refers to kilocalorie)
Malnutrition	Deficiency or excess of nutrients
Macronutrient	Nutrients needed in large amounts (carbs, fats, proteins)
Micronutrient	Nutrients needed in small amounts (vitamins, minerals)

Calculation Examples on Energy and Nutrition

Example 1: A patient consumes a meal that contains 30 grams of carbohydrates, 15 grams of protein, and 10 grams of fat. How many total calories does this meal provide? Calculate the energy contribution from each macronutrient and determine the overall caloric content of the meal.

Solution:

- Calories from carbohydrates = 30 g × 4 kcal/g = 120 kcal
- Calories from protein = 15 g × 4 kcal/g = 60 kcal
- Calories from fat = 10 g × 9 kcal/g = 90 kcal

Total calories = 120 + 60 + 90 = 270 kcal

Example 2: A patient requires 1.2 grams of protein for each kilogram of body weight. If the patient's weight is 70 kilograms, calculate the total amount of protein needed per day to meet their nutritional requirement.

Solution: Protein requirement = 1.2 g × 70 kg = 84 grams of protein per day

Check Your Self

A 6-year-old malnourished child consumes very small amounts of food each day. His typical daily intake includes one slice of bread (containing 12 grams of carbohydrates, 2 grams of protein, and 1 gram of fat), one boiled egg (providing 6 grams of protein and 5 grams of fat), and half a cup of sweetened tea with 5 grams of sugar. Calculate the total caloric intake from this food using standard nutritional values (4 kcal/g for carbohydrates and protein, 9 kcal/g for fat), and evaluate whether this amount is sufficient for a growing child.(The normal level usually ranges between 1200 to 1600 calories depending on the activity level).

Chemistry Link to Health

AI and Biomedical Measurements

Accurate measurement of chemical parameters such as pH, pressure, density, and energy metabolism plays a vital role in effective clinical care. With the rapid advancement of technology, these measurements are increasingly integrated with Artificial Intelligence (AI) to improve diagnostic accuracy and therapeutic outcomes. Recent studies have highlighted the growing use of AI in biomedical applications, particularly in interpreting biochemical data to detect acid-base imbalances like acidosis and alkalosis. AI algorithms can analyze complex datasets with greater speed and precision than conventional methods, enabling healthcare providers to make timely and informed decisions.

Looking ahead, AI is poised to transform the landscape of clinical practice through several promising developments. **Continuous monitoring** using smart biosensors will allow real-time tracking of biochemical parameters, supporting more dynamic and responsive care. **Early diagnosis** will be enhanced through predictive analytics, enabling clinicians to identify potential health issues before symptoms become evident. Moreover, **personalized treatment** strategies will become more feasible, as AI systems analyze individual biochemical profiles to tailor therapies that are more effective and patient-specific. This integration of chemistry, technology, and artificial intelligence marks a significant step toward more precise, proactive, an personalized healthcare.

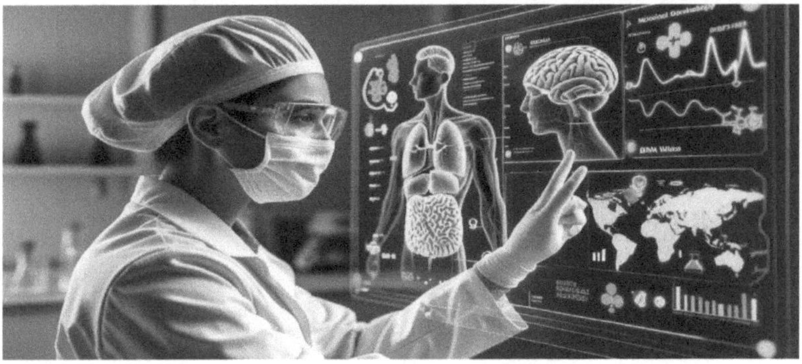

Chapter summary

This chapter introduced the core principles of chemical measurements and their application in nursing and health sciences. It began with a review of the standard units of measurement used in clinical practice, focusing on the metric and U.S. customary systems, along with essential techniques for accurate unit conversion.

Key measurement concepts such as **significant figures**, **scientific notation**, **accuracy**, **precision**, and **percent error** were explored to highlight their importance in safe medication administration, laboratory result interpretation, and clinical decision-making.

The chapter also examined the biomedical relevance of **density**, **temperature**, and **pressure**, demonstrating how these physical properties influence diagnostics, fluid management, and physiological monitoring in patient care.

A dedicated section on **pH and acid–base balance** explained how hydrogen ion concentration affects the body's internal environment. The clinical significance of maintaining normal blood pH (7.35–

7.45) was emphasized, along with methods for calculating pH and identifying conditions like acidosis and alkalosis.

Finally, the chapter concluded with an overview of **energy and nutrition**, outlining how the body derives energy from macronutrients and highlighting the role of metabolism in supporting health and recovery.

Together, these topics provide nursing students with a strong foundation in chemical principles essential for accurate, safe, and effective clinical practice.

EXERCISES

Review Questions

1. Why is it important to use standard units of measurement in biomedical settings?
2. How does density affect the behavior of fluids in the human body?
3. Describe the relationship between temperature and metabolic activity.
4. What role does pressure play in the cardiovascular and respiratory systems?
5. How do acid-base imbalances affect human health? Give an example of a condition caused by pH disturbances.
6. Why is accurate temperature measurement critical in clinical diagnostics?
7. Why is understanding energy metabolism important for patient nutrition management?

Calculation Problems

1. A patient's blood volume is 5 liters, and the density of blood is approximately 1.06 g/mL. Calculate the mass of the patient's blood in kilograms.
   (Hint: 1 liter = 1000 mL)
2. Convert 98.6°F (normal body temperature) to Celsius using the formula:
   $°C = (°F - 32) \times 5/9$

3. A nurse needs to prepare 500 mL of a saline solution at pH 7.4. If the solution's pH changes to 7.0, what is the difference in hydrogen ion concentration? *(Hint: $pH = -log[H^+]$)*
4. A patient receives a parenteral nutrition bag with 80g of glucose. Calculate the total kilocalories provided from glucose. *(Hint: 4 kcal/g)*

Fill in the Blanks

1. The basic unit of pressure in the metric system is the ________.
2. The normal pH range of human blood is between ________ and ________.
3. Energy in food is measured in units called ________.
4. The formula to convert Celsius to Fahrenheit is ________.

Multiple Choice

1. Which unit is **NOT** part of the metric system?
   A) Meter
   B) Gram
   C) Pound
   D) Liter
2. The condition where blood pH is below 7.35 is called:
   A) Alkalosis
   B) Acidosis
   C) Neutralization
   D) Homeostasis
3. Which of the following best describes density?
   A) The amount of energy in a substance
   B) The volume of fluid per unit of time
   C) The mass of a substance per unit volume
   D) The rate of temperature change

True or False

1. The American unit for pressure is the pascal. (True / False)
2. Density is defined as mass divided by volume. (True / False)

3. Blood pressure is measured in millimeters of mercury (mmHg). (True / False)
4. Acid-base balance has no effect on enzyme function in the body. (True / False)

Basic Calculations

1. A meal contains 50g carbs, 25g protein, and 15g fat. Calculate total kcal.
2. Convert 120g of carbohydrates into kilocalories.
3. A patient drinks a beverage with 20g alcohol. How many kcal does it provide? *(Hint: 7 kcal/g)*
4. Convert 2,500 kcal to kilojoules (kJ). *(Hint: 1 kcal = 4.184 kJ)*
5. Calculate total kcal in: 30g protein + 10g fat + 5g alcohol.

Clinical Applications

1. A diabetic patient's snack has 40g carbs, 10g protein, and 12g fat. Is this within a 300-kcal limit?
2. A 75kg patient needs 1.8g protein/kg/day. How much protein (in grams) must they consume?
3. An IV solution provides 500 kcal via 100g dextrose (carbs). How many grams of fat would yield the same energy? *(Hint: 9 kcal/g)*

Advanced Problems

1. **Post-surgery nutrition:**
   - Target: 2,000 kcal/day
   - 55% from carbs, 25% from protein, 20% from fat.
   - Calculate grams of each macronutrient.
2. **Alcohol substitution:**
   - A cocktail has 150 kcal from alcohol.
   - If replaced with fat (same kcal), how many grams of fat are needed?

Chapter 3

Chemical Bonds and Matter in Healthcare

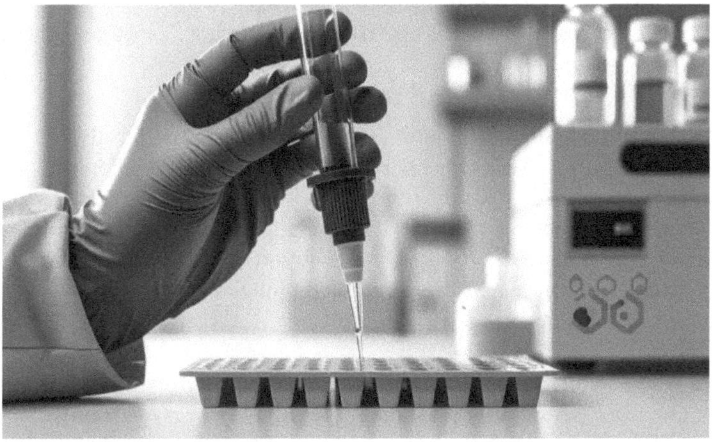

Chapter 3: Chemical Bonds and Matter in Healthcare

Learning Objectives

By the end of this chapter, you should be able to:

1. Differentiate between elements, compounds, and mixtures.
2. Describe the structure of atoms and how they relate to matter.
3. Distinguish between ionic and covalent bonds.
4. Explain how chemical bonds contribute to molecular stability and function.
5. Interpret the role of chemical compounds in clinical practice and patient care

Understanding the basic nature of matter is essential for all healthcare professionals, as every structure and substance in the human body—cells, tissues, fluids, and medications—is composed of matter. This chapter introduces the fundamental chemical principles that explain how matter is structured, how it behaves under different conditions, and how it interacts at the molecular level within biological systems.

We begin by differentiating between the forms of matter—elements, compounds, and mixtures—and then explore how atoms form chemical bonds to create the complex molecules that sustain life. A strong understanding of chemical bonding, particularly ionic and covalent interactions, is crucial for explaining essential physiological processes such as drug activity, nutrient transport, and fluid-electrolyte balance.

Importantly, this chapter links chemical theory to real-world clinical applications. By the end of the chapter, students will be able to relate the structure and stability of compounds to their biological roles in diagnostics, treatments, and metabolic functions. From how antacids neutralize stomach acid to the formulation of intravenous fluids, the chemistry of matter and bonding is a foundation of safe and effective nursing care.

Atoms and Elements

Atoms are the smallest units of matter that retain the identity of an element, and they serve as the basic building blocks of everything in the human body and clinical environment. From the oxygen we inhale to the calcium in our bones, atoms of elements are involved in nearly every physiological function and medical procedure.

An **element** is a pure substance consisting of only one type of atom. Of the more than 100 known elements, approximately 20 are essential to human health. Elements such as carbon (C), hydrogen (H), oxygen (O), and nitrogen (N) make up most of the body's mass, while others like iron (Fe), calcium (Ca), sodium (Na), and iodine (I) are required in smaller amounts to support vital processes like oxygen transport, nerve signaling, and hormone regulation.

A solid understanding of atoms and elements is crucial for interpreting laboratory results, managing electrolyte imbalances, and safely administering medications. This section introduces the structure of atoms, the periodic table of elements, and the biomedical significance of key elements in health and disease.

Structure of the Atom

Atoms are composed of three fundamental subatomic particles: **protons**, **neutrons**, and **electrons**. At the center of the atom is the **nucleus**, which contains **protons**—positively charged particles—and **neutrons**, which have no charge. Surrounding the nucleus are **electrons**, negatively charged particles that move in regions called **energy levels** or **electron shells**. Although electrons are much smaller in mass than protons and neutrons, their arrangement determines how atoms interact with one another to form chemical bonds.

In healthcare, understanding the structure of the atom is vital, as it helps explain how ions are formed, how electrolytes function in the body, and how certain diagnostic tools—like PET scans—rely on the behavior of subatomic particles.

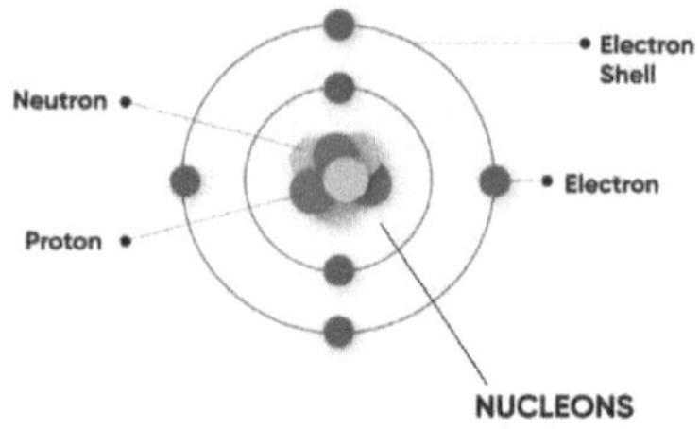

Figure 3.1 Model of the atom

Atomic Number and Mass Number

The **atomic number** of an element is the number of **protons** in the nucleus of its atoms. It defines the element's identity because every atom of that element contains the same number of protons. For example, oxygen has an atomic number of 8, meaning each oxygen atom contains 8 protons.

The **mass number** represents the total number of **protons and neutrons** in the nucleus of an atom. Since neutrons have nearly the same mass as protons but no charge, they contribute significantly to the atom's overall mass. For example, a typical carbon atom has 6 protons and 6 neutrons, so its mass number is 12.

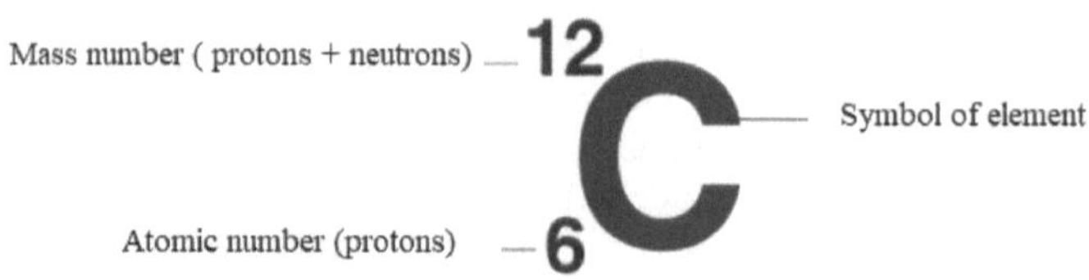

Figure 3.2 Atomic Structure: Carbon as an Example

Together, the atomic number and mass number help describe different forms of an element called **isotopes**, which have the same number of protons but different numbers of neutrons. Understanding these concepts is important in healthcare, especially in medical imaging and radiation therapy, where specific isotopes of elements like iodine or technetium are used.

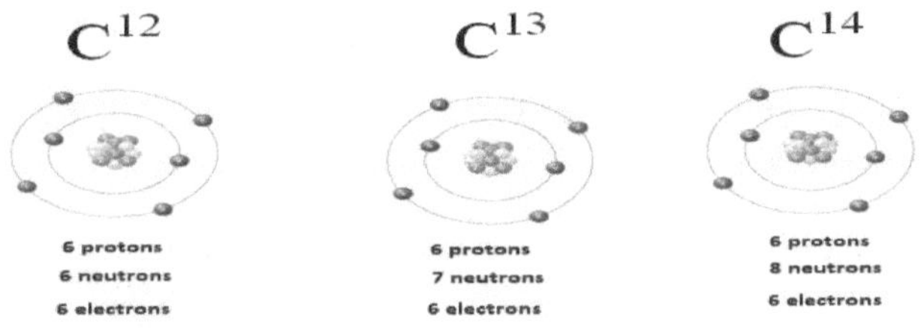

Figure 3.3 Isotopes of Carbon

Isotopes and Their Medical Use

Isotopes are atoms of the same element that have the same number of protons but different numbers of neutrons, giving them different mass numbers. Because they have the same number of electrons, isotopes of an element generally behave the same in chemical reactions. However, some isotopes are unstable and release energy as they decay—these are called radioisotopes.

Nuclear Chemistry in Healthcare

Radioisotopes form the foundation of several essential diagnostic and therapeutic tools in modern medicine. In nuclear imaging, isotopes such as technetium-99m are commonly used with gamma cameras and SPECT (Single Photon Emission Computed Tomography) scanners to visualize the heart, bones, and kidneys. Positron-emitting isotopes like fluorine-18 are used in PET (Positron Emission Tomography) scans to assess metabolic activity, especially in cancer and brain disorders.

For treatment, isotopes such as iodine-131 and cobalt-60 emit beta and gamma radiation used to selectively target diseased tissues—especially in thyroid disorders and external beam radiotherapy for cancer. Linear accelerators (LINACs) are used to generate high-energy X-rays or electrons for precise tumor targeting.

Nurses working in nuclear medicine, oncology, and radiology departments must understand the basics of nuclear chemistry to ensure radiation safety, assist with radiopharmaceutical administration, operate or support the use of shielding equipment, and educate patients on pre- and post-procedure precautions.

Radiation types encountered in healthcare include alpha particles (short-range, high-energy), beta particles (used in localized treatments), gamma rays (for deep tissue penetration), and positrons (used in PET scans).

In addition, some sterilization systems use gamma radiation from cobalt-60 to disinfect heat-sensitive medical tools, ensuring high standards of infection control. As nuclear chemistry advances, its role in personalized medicine and minimally invasive diagnostics continues to grow—making it an essential area of knowledge for healthcare professionals.

In healthcare, radioisotopes have important diagnostic and therapeutic uses. For example:

- **Iodine-131** is a radioactive isotope used to diagnose and treat **thyroid disorders**, including hyperthyroidism and thyroid cancer, because iodine naturally accumulates in the thyroid gland.
- **Technetium-99m** is widely used in **nuclear medicine imaging** (such as bone scans and cardiac stress tests) because it emits gamma rays that can be detected by special cameras without causing much damage to tissues.
- **Cobalt-60** is used in **radiation therapy** to target and destroy cancer cells.

Understanding isotopes allows healthcare professionals to better interpret diagnostic imaging results and monitor or assist in treatments involving radiation.

Clinical Case Example

Patient Scenario: A 45-year-old woman presents with symptoms of weight loss, anxiety, and a rapid heartbeat. Her physician suspects **hyperthyroidism** and orders a **radioactive iodine uptake (RAIU)** test.

Application of Isotopes: The patient receives a small oral dose of **iodine-131 (^{131}I)**, a radioactive isotope that accumulates in the thyroid gland. A **gamma camera** is then used to measure how much of the isotope the thyroid absorbs. **Increased uptake** indicates **overactive thyroid tissue**, confirming the diagnosis. If necessary, a

therapeutic dose of iodine-131 may be given later to selectively **destroy excess thyroid tissue**.

Why It Matters: This case highlights the critical importance of understanding isotopes in clinical care. Nurses play a central role in safely managing radioactive treatments, closely monitoring patients for potential side effects such as dry mouth, neck tenderness, or fatigue, and providing clear education on both pre-test preparation and post-treatment radiation precautions, including the need to avoid close contact with others for a specified period. Such knowledge enables nurses to ensure patient safety, support therapeutic outcomes, and minimize radiation-related risks.

The following table summarizes some of the most commonly used isotopes in medical practice.

Table 3.1 Comparison Common Isotopes in Medicine

Isotope	Element	Mass Number	Medical Use	Mode of Action
Iodine-131	Iodine (I)	131	Treats hyperthyroidism & thyroid cancer	Accumulates in thyroid tissue and emits beta radiation
Technetium-99	Technetium (Tc)	99	Diagnostic imaging (e.g. bone scans, cardiac scans)	Emits gamma rays for imaging without high tissue damage
Cobalt-60	Cobalt (Co)	60	Radiation therapy for cancer	Emits gamma radiation to destroy cancer cells
Carbon-14	Carbon (C)	14	Metabolic and research studies	Tracer in biochemical reactions

Sources and Types of Radiation in Healthcare

In medical practice, radiation mainly comes from **radioactive isotopes** or from **medical devices** that generate radiation. Radioactive isotopes are unstable atoms that release energy in the form of radiation as they break down over time. These isotopes can be produced in nuclear reactors or particle accelerators and are used in small, controlled amounts for diagnosis and treatment. Medical devices, such as X-ray machines or linear accelerators, produce radiation without using radioactive materials.

There are several types of radiation commonly used in healthcare, and each has specific properties and applications:

Alpha radiation (α) consists of heavy particles made of two protons and two neutrons. It cannot penetrate the skin but can be dangerous if inhaled or swallowed. Alpha particles are used in targeted treatments for bone cancer, such as radium-223 therapy.

Beta radiation (β) is made of fast-moving electrons or positrons. It can penetrate a few millimeters into the body and is used both for treatment and imaging. A well-known example is iodine-131, used to treat thyroid conditions. $Beta^{+}$ particles are also used in PET scans, where they help detect tumors and assess organ function.

Gamma radiation (γ) is a form of high-energy light that can pass through the body. It is widely used in nuclear imaging (such as bone or thyroid scans) and in radiotherapy to destroy cancer cells. Gamma rays are also used to sterilize medical tools and equipment.

X-rays are similar to gamma rays but are produced by machines. They are used in many diagnostic tests, including chest X-rays, dental exams, and CT scans. High-energy X-rays from linear accelerators are also used in cancer treatment.

Positrons, the positively charged counterparts of electrons, are used in PET imaging. When positrons collide with electrons in the body, they produce gamma rays that help create detailed images of organs and tissues.

Why This Matters in Nursing

As a nurse, having a solid understanding of radiation is crucial because it directly impacts both patient safety and your own well-being when working in clinical settings involving diagnostic imaging or cancer treatment. Patients undergoing procedures such as X-rays, CT scans, nuclear medicine studies, or radiation therapy are exposed to different types of radiation, each with specific effects and risks. Your role may include preparing and administering radioactive medications, handling radioactive materials with care, and ensuring that proper safety protocols—such as the use of shielding devices and adherence to time and distance principles—are rigorously followed to minimize unnecessary exposure.

Moreover, you play a key role in educating patients and their families about what to expect during radiation procedures, addressing concerns about potential side effects, and providing guidance on safety measures to protect themselves and others after treatment. Understanding radiation biology also helps you recognize early signs of radiation-related complications, enabling prompt interventions that can improve patient outcomes.

In addition, knowing the principles of radiation safety helps you advocate for safe workplace practices, comply with regulatory standards, and participate effectively in multidisciplinary care teams. Ultimately, mastering radiation knowledge empowers you to deliver care that is not only effective but also prioritizes the safety and dignity of both patients and healthcare professionals.

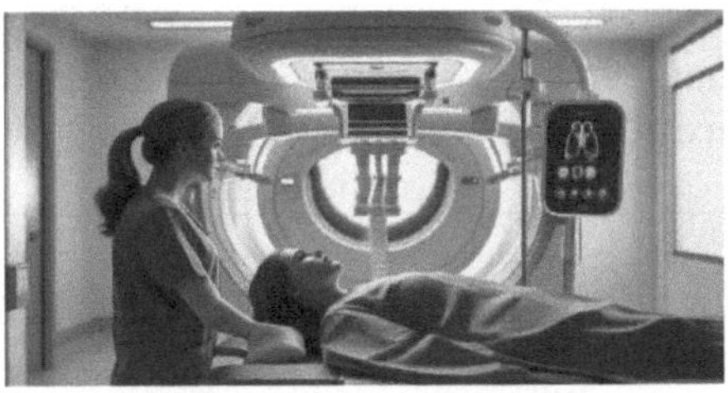

Figure 3.4 Radiation Therapy: Precision Use of High-Energy X-rays in Cancer Treatment

The following table summarizes the main types of radiation used in healthcare, along with their medical uses and associated tools or devices.

Table 3.2 Types of Medical Radiation, Applications, and Devices

Radiation Type	Description	Clinical Applications	Devices / Tools
Alpha (α)	Heavy particles; low penetration	Targeted cancer therapy (e.g., Radium-223)	Sealed alpha-emitting sources in radiotherapy units
Beta (β^-/β^+)	High-speed electrons or positrons	Thyroid treatment (I-131), PET scans	Radiopharmaceuticals, PET scanners
Gamma (γ)	High-energy waves; deep penetration	Nuclear imaging, radiotherapy, sterilization	Gamma cameras, Cobalt-60 therapy machines
X-rays	Artificial electromagnetic radiation	Imaging (chest, bone), CT, radiotherapy	X-ray machines, CT scanners, LINACs
Positrons (β^+)	Positively charged electrons	PET imaging (tumors, brain, heart)	PET scanners, cyclotrons

Chemical Bonds

In the human body and in pharmaceutical compounds, atoms seldom exist in isolation. Instead, they combine to form molecules and compounds through **chemical bonds**, which are essential to biological structure and function. A **chemical bond** is the attractive force that holds atoms together within a compound, and the nature of this bond depends on how electrons are distributed—either transferred or shared—between atoms.

There are two primary types of chemical bonds relevant to biology and medicine: **ionic bonds** and **covalent bonds**.

Ionic Bonds

An **ionic bond** is formed when one atom **transfers electrons** to another, resulting in the formation of charged particles known as **ions**. The atom that **loses electrons** becomes a **positively charged ion (cation)**, while the atom that **gains electrons** becomes a **negatively charged ion (anion)**. These oppositely charged ions are held together by **strong electrostatic attraction**, forming an ionic bond. This type of bonding is common in compounds like **sodium chloride (NaCl)**, and it plays a crucial role in physiological processes such as nerve signal transmission and fluid balance.

Ionic bonds typically form between a **metal** and a **nonmetal**. For example:

- **Sodium (Na)** is a metal that easily **loses one electron**, becoming Na^+.
- **Chlorine (Cl)** is a nonmetal that readily **gains one electron**, becoming Cl^-.
- Together, they form **sodium chloride (NaCl)**, or table salt, which dissolves in water to form **electrolytes**.

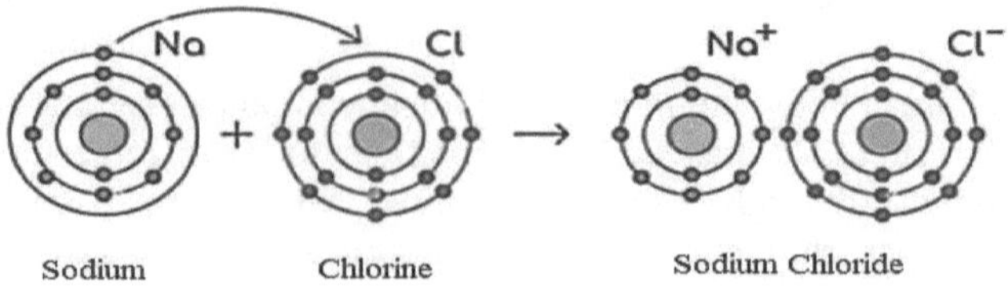

Figure 3.5 Ionic bond formation in Sodium Chloride

Clinical Connection: Electrolytes in the Body

When **ionic compounds** such as **sodium chloride (NaCl)** dissolve in body fluids, they separate into charged particles called **ions.** Common biologically important ions include:

- **Sodium (Na^+)**
- **Potassium (K^+)**
- **Chloride (Cl^-)**
- **Calcium (Ca^{2+})**

These **electrolytes** play essential roles in maintaining physiological balance and are involved in:

- **Nerve impulse transmission** – Sodium and potassium are crucial for generating and propagating electrical signals in neurons.
- **Muscle contraction** – Calcium triggers contraction in skeletal, cardiac, and smooth muscle tissues.
- **Heart rhythm regulation** – Electrolyte imbalances can lead to arrhythmias and cardiac complications.
- **Fluid and pH balance** – Electrolytes regulate osmotic pressure, hydration, and acid-base equilibrium.

Nursing Relevance:

Monitoring and correcting electrolyte levels is a routine yet critical part of nursing care. Nurses use **blood tests** to assess electrolyte concentrations and respond by administering **IV fluids, oral**

supplements, or adjusting **medications**. Rapid intervention may be life-saving in cases of **dehydration**, **renal dysfunction**, **burns**, or **postoperative fluid shifts**.

Covalent Bonds

A **covalent bond** forms when **two atoms share one or more pairs of electrons** to achieve stability. Unlike ionic bonds (which involve the transfer of electrons), covalent bonds usually occur between **nonmetallic elements** and create **molecules** rather than ionic compounds.

In a covalent bond, atoms share electrons to complete their **outer electron shells**, allowing each atom to achieve a stable configuration—often resembling the electron arrangement of noble gases.

There are different types of covalent bonds:

- **Single bond**: One pair of shared electrons (e.g., H–H in H_2)
- **Double bond**: Two pairs of shared electrons (e.g., O=O in O_2)
- **Triple bond**: Three pairs of shared electrons (e.g., N≡N in N_2)

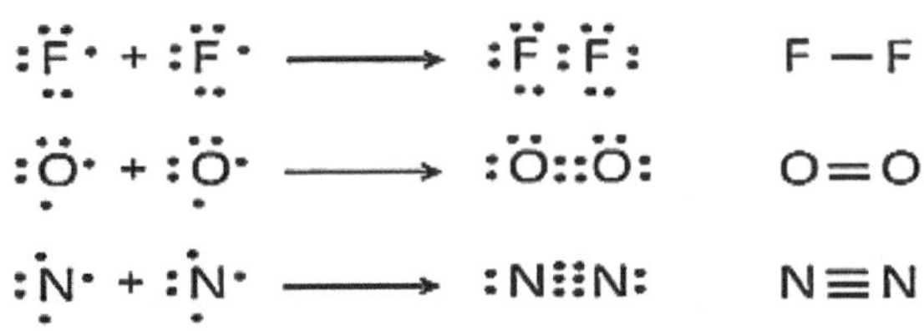

Figure 3.6 covalent bond formation

Table 3.3 Types of Covalent Bonds

Bond Type	Number of Shared Electron Pairs	Example	Description
Single bond	1 pair (2 electrons)	H–H (Hydrogen gas)	Each atom shares one electron; simplest covalent bond.
Double bond	2 pairs (4 electrons)	O=O (Oxygen gas)	Atoms share two pairs of electrons; stronger and shorter than single bonds.
Triple bond	3 pairs (6 electrons)	N≡N (Nitrogen gas)	Atoms share three pairs; very strong and short bond.

Healthcare Connection: Covalent Bonds in the Body

Many essential biomolecules in the human body—such as **proteins, carbohydrates, lipids, DNA, and medications**—are composed of atoms held together by **covalent bonds**.

Examples:

- **Water (H_2O)** is formed by covalent bonds between hydrogen and oxygen atoms. It is essential for virtually all physiological processes.
- **Oxygen gas (O_2)**, required for cellular respiration, consists of two oxygen atoms connected by a **double covalent bond**.
- **Medications** often form covalent or weak polar interactions with specific **receptors** to produce their therapeutic effects.

Nursing Relevance:
Understanding covalent bonds helps nurses grasp how medications are structured, how they interact with receptors, and why **solubility**, **absorption**, and **metabolism** depend on molecular structure.

Other Types of Bonds and Interactions in Healthcare

In addition to covalent and ionic bonds, many **weak interactions** play essential roles in the structure and function of biological molecules. These include **hydrogen bonds**, **Van der Waals forces**, **dipole–dipole interactions**, **disulfide bonds**, and **ionic attractions** within proteins. Although individually weak, these bonds collectively stabilize complex structures like **proteins**, **enzymes**, and **DNA**, and govern how **drugs bind to their receptors**.

In clinical settings, these interactions help explain how medications function, how protein shape affects hormone activity, and how physiological changes (like pH shifts or fever) can disrupt critical biological processes. For nurses, understanding these subtle yet powerful forces is vital for interpreting lab results, managing drug therapy, and recognizing the molecular basis of disease.

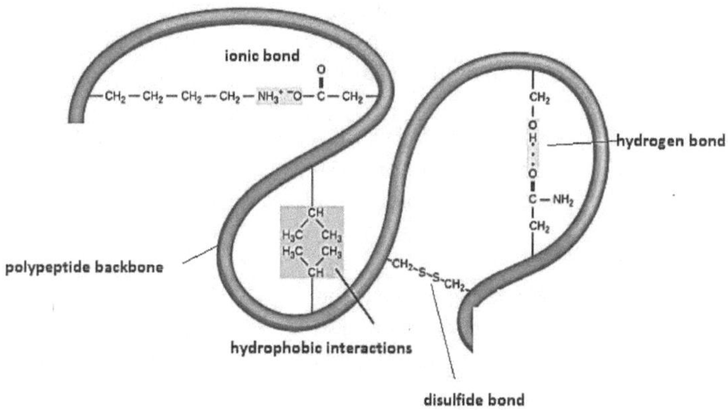

Figure 3.7 Bond Interactions that determine a protein's tertiary structure

Table 3.4 summarizes the key types of non-covalent bonds and interactions commonly found in biological systems and highlights their significance in healthcare and nursing practice.

Table 3.4 Other Types of Bonds and Interactions in Healthcare

Bond / Interaction	Description	Example in the Body or Healthcare
Hydrogen Bond	A weak attraction between a hydrogen atom (already bonded to O or N) and another electronegative atom (like O or N).	Stabilizes the double-helix structure of DNA and helps proteins fold properly.
Van der Waals Forces	Weak, short-range forces due to temporary shifts in electron clouds.	Help molecules interact temporarily, such as in enzyme-substrate binding.
Dipole–Dipole Interactions	Attractions between molecules with permanent partial charges (dipoles).	Found in polar molecules like water and some drug molecules.
Disulfide Bond	A covalent bond between two sulfur atoms (usually in cysteine amino acids).	Helps maintain protein structure, especially in hormones like insulin.
Ionic Interactions (in proteins)	Weak ionic attractions between charged side chains in amino acids.	Important in protein folding and receptor binding.

Why These Bonds Matter in Nursing

Understanding weak chemical interactions is essential for nurses because these forces play a central role in many clinical and physiological processes:

1. **Drug–Receptor Binding**: Medications often rely on **hydrogen bonds** and **van der Waals forces** to interact precisely with their

biological targets. Proper binding ensures therapeutic efficacy and minimizes side effects.

2. **Protein Structure and Function**: Weak interactions such as **ionic attractions** and **dipole–dipole forces** are crucial for maintaining a protein's three-dimensional shape. Disruption due to changes in **pH**, **temperature**, or **electrolyte balance**—common in many clinical conditions—can impair protein function.
3. **Hormone Stability**: For example, the biological activity of **insulin** depends on **disulfide bonds** that maintain its active conformation. Disruption of these bonds may render the hormone ineffective.

States of Matter and Bonds

Matter exists in **three primary states** under normal physiological conditions: **solid**, **liquid**, and **gas**. The **type and strength of bonds** between particles influence each state's physical properties, which is crucial for understanding how medications are stored, administered, and metabolized.

Solids consist of particles packed tightly in a fixed, ordered structure. These particles are held together by **strong intermolecular forces**, which can be ionic bonds (as in table salt, NaCl), covalent bonds, or hydrogen bonding. Consequently, solids exhibit **definite shape and definite volume** due to the particles' inability to flow past one another. The particles possess **very low kinetic energy**, primarily vibrating in place rather than moving freely. This combination of rigid structure and strong bonding results in materials that are stable and provide structural integrity, for instance, **table salt (NaCl)** maintained by its ionic lattice, or **bone mineral (calcium phosphate)**, whose stable solid form offers crucial support to skeletal systems.

Liquids

Liquids consist of particles that remain **close together** but possess sufficient energy to **move past one another**. They are held by **moderate intermolecular forces**, such as hydrogen bonds or dipole-dipole interactions. This balance between cohesion and mobility gives liquids a **definite volume** but allows them to **take the shape of their container**, flowing to conform to its boundaries. A prime example is **water (H_2O)**, where hydrogen bonding underpins its unique properties essential for biological processes. Another key example is **blood plasma**, a complex fluid whose liquid nature enables it to transport dissolved ions, proteins, and cells throughout the circulatory system.

Gases are characterized by particles that are **far apart** and experience **minimal attraction** due to **very weak or negligible intermolecular forces**. This separation and lack of strong bonding allow the particles to move rapidly and randomly with **very high kinetic energy**. Consequently, gases have **no fixed shape or volume**; they expand completely to fill any container they occupy. Common examples include essential respiratory gases like **oxygen (O_2) and carbon dioxide (CO_2)**, which diffuse freely within the lungs and bloodstream, and **inhaled anesthetics**, whose gaseous state allows precise delivery and rapid distribution within the body during medical procedures.

Plasma, the fourth state of matter constituting over 99% of the visible universe (e.g., stars, lightning), forms when **extremely high energy** strips electrons from atoms, creating a **highly energized mixture of free electrons and positive ions**. Unlike other states, plasma has **no fixed shape or volume** and expands like a gas, while its charged particles enable **efficient electricity conduction** and **strong response to magnetic fields**. These properties make it transformative in healthcare: **Cold plasma sterilization** leverages its reactive ions to disinfect heat-sensitive instruments (like endoscopes/catheters), killing pathogens without thermal damage, while **plasma arc surgical tools** use focused ionized gas to cut tissue and coagulate blood with precision, minimizing bleeding and collateral damage. For nurses, understanding these applications ensures safe handling of plasma-sterilized equipment and supports infection control in surgical settings. Figure 3.8 States of Matter

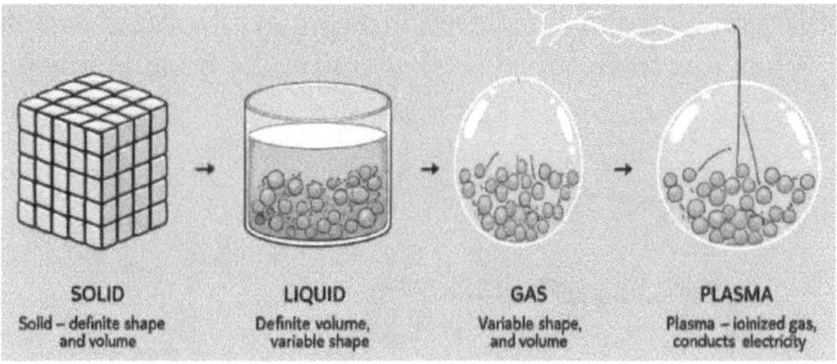

Figure 3.8 States of Matter

Table 3.4 States of Mater and Healthcare

State	Healthcare Example	Importance
Solid	Tablets, bone tissue	Structure, controlled drug release
Liquid	Blood, IV fluids	Transport, hydration, medication delivery
Gas	Oxygen, anesthesia	Respiratory care, sedation
Plasma	Sterilization systems	Infection control, surgical applications

Polarity and Solubility

In healthcare and biology, understanding **polarity** is key to predicting whether a substance will **dissolve in water**, be absorbed by the body, or mix with other medications.

Polarity

Polarity describes how electrons are shared in a chemical bond and how this affects the **distribution of electrical charge** in a molecule.

What Causes Polarity?

Polarity arises when atoms in a molecule have **different electronegativities** (the ability to attract electrons).

- When electrons are shared **unequally**, the bond is **polar**.
- When electrons are shared **equally**, the bond is **nonpolar**.

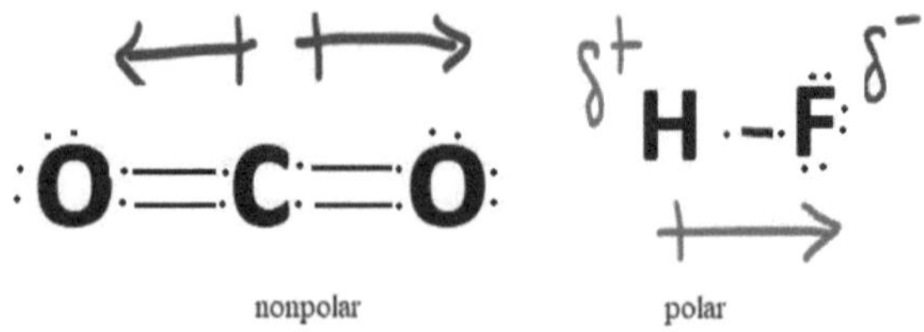

Figure 3.9 Polar and nonpolar molecules

Polar molecules have a **partial positive charge** on one side and a **partial negative charge** on the other—forming a **dipole**.

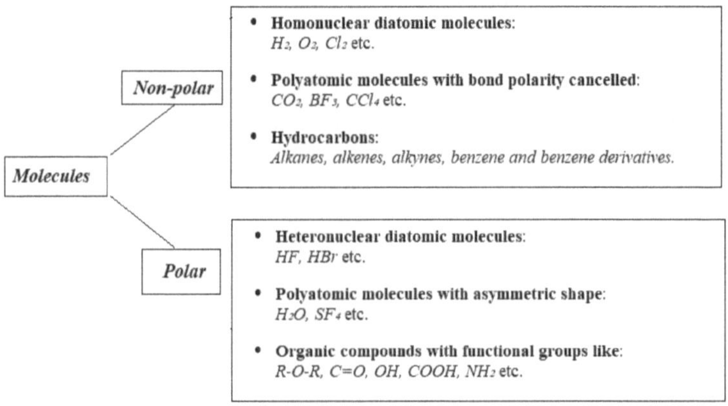

Figure 3.10 Examples of polar and nonpolar molecules

Polarity and Molecular Behavior

Polarity critically influences molecular behavior, primarily through strong intermolecular forces like hydrogen bonding and dipole-dipole interactions. Polar molecules exhibit high solubility in water because their charged regions form favorable electrostatic attractions with water's polar molecules, allowing effective dissolution (e.g., ethanol mixing readily with water). Conversely, these same strong intermolecular forces require more energy to overcome, resulting in higher boiling and melting points for polar substances compared to nonpolar ones of similar size. For instance, water (H_2O, polar) boils at 100°C, while methane (CH_4, nonpolar) boils at -161°C, illustrating how polarity dramatically increases the energy needed to transition phases.

"Like Dissolves Like": The Molecular Basis of Solubility

The phrase "like dissolves like" means that polar substances dissolve in polar solvents (like water), while nonpolar substances dissolve in nonpolar solvents (like oils). This principle is based on the compatibility of intermolecular forces. Polar molecules form strong dipole–dipole or hydrogen bonds, which can only be replaced by similar interactions in polar solvents. Nonpolar molecules rely on weaker dispersion forces, best preserved in nonpolar environments. Thermodynamically, dissolution is favored when it increases entropy and forms energetically stable interactions. Therefore, solubility depends on both molecular similarity and energy balance.

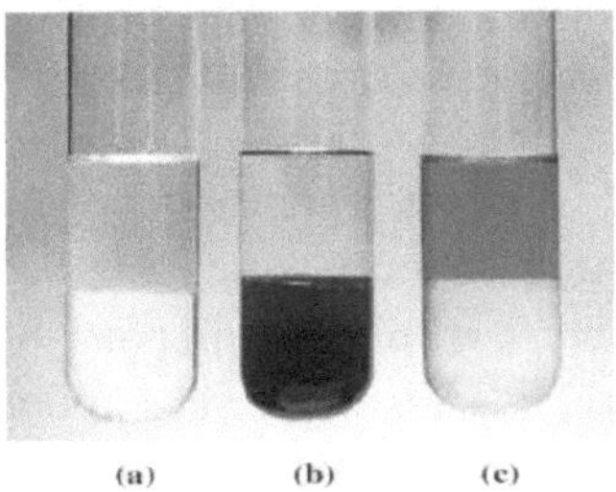

Figure 3.11 Like dissolves like

In each test tube, the lower layer is CH_2Cl_2 (more dense), and the upper layer is water (less dense). (a) CH_2Cl_2 is nonpolar and water is polar; the two layers do not mix. (b) The nonpolar solute I2 (purple) is soluble in the nonpolar CH_2Cl_2 layer. (c) The ionic solute Ni $(NO_3)_2$ (green) is soluble in the polar water layer.

Polarity and Biological Interactions

Understanding the polarity of drugs is essential for predicting their behavior within the human body, particularly in clinical settings. Polarity—defined by how unevenly electrical charges are distributed across a molecule—affects multiple aspects of drug action, including absorption, distribution, metabolism, and excretion. Polar (water-soluble) drugs dissolve readily in the aqueous environment of blood and are often excreted quickly by the kidneys. In contrast, nonpolar (fat-soluble) drugs tend to accumulate in fatty tissues, where they can be stored for longer durations and released slowly over time.

The polarity of a drug also influences its ability to cross biological barriers. Cell membranes are primarily composed of nonpolar lipids, meaning that nonpolar drugs can diffuse across them freely, while polar drugs require specific transport proteins or channels to enter or exit cells. This has critical implications for drug delivery and therapeutic effectiveness. Furthermore, polarity plays a key role in intravenous (IV) drug compatibility. Mixing polar and nonpolar substances in IV solutions can lead to drug precipitation or inactivation, posing risks to patient safety.

Nursing Tip: Before administering medications—especially intravenously—nurses must consider the polarity of each drug. Knowing whether a medication is polar or nonpolar helps prevent incompatibilities, ensures proper absorption, and avoids potentially harmful interactions in IV lines. The following table outlines common examples of how drug polarity affects solubility, storage, membrane passage, and clinical use.

Table 3.5 Polarity and Drug Behavior in the Body

Concept	Example
Water-soluble drugs	Dissolve easily in blood and are excreted faster (e.g., aspirin, penicillin, vitamins B and C).
Fat-soluble drugs	Absorbed into body fat and stored longer (e.g., vitamins A, D, E, K).
Drug interactions	Some medications precipitate if mixed due to polarity mismatch.
Cell membranes	Made of lipids — nonpolar drugs can pass freely, polar ones need transport.

Chemistry Link to Health

Smart Solvents and Drug Delivery: A New Era in Polarity and Bonding

A groundbreaking 2024 study from the University of Texas at Austin has pioneered adaptive "smart solvents" — molecular cages stabilized by *covalent bonds* that dynamically shift their *polarity* in response to environmental triggers (e.g., pH, temperature). These cages exploit *hydrogen bonding* and *dipole interactions* to encapsulate drugs (including ionic salts like morphine sulfate), adhering to the *"like dissolves like"* principle to optimize solubility for polar or nonpolar compounds. By releasing drugs only at target sites, they overcome *poor solubility* — a key cause of clinical trial failures — enabling precise delivery with fewer side effects, enhanced absorption of lipid-soluble agents (e.g., vitamins A/D/E/K), and safer formulations using biodegradable solvents. For nurses, this technology underscores the critical role of solubility principles: it informs IV compatibility assessments, guides administration of fat-soluble drugs with meals, and sharpens monitoring for solvent-related adverse reactions, while advancing *sustainable healthcare* by reducing pharmaceutical waste.

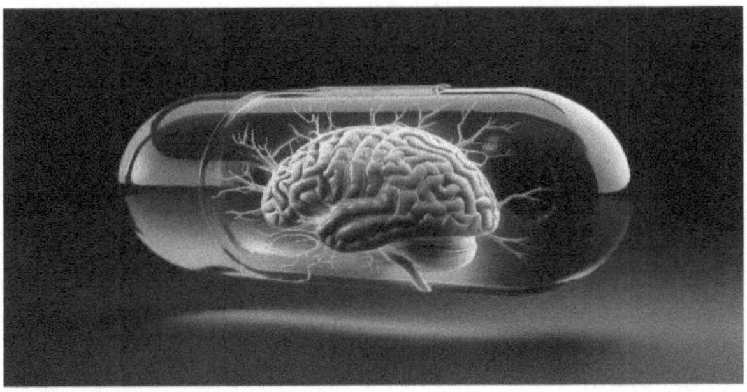

Chapter Summary

All matter—whether it supports life through electrolytes or powers advanced medical technologies—is composed of atoms. These atoms, built from protons, neutrons, and electrons, are defined by their atomic number. Differences in neutron number result in isotopes, some of which are critical in medicine, such as iodine-131 used in thyroid imaging. Atomic stability is achieved through chemical bonding: ionic bonds (involving electron transfer, as in NaCl), covalent bonds (involving electron sharing, as in H_2O), and hydrogen bonds. These interactions influence molecular properties, particularly polarity. Polar molecules (like water) dissolve other polar substances, while nonpolar compounds (like fats) require nonpolar solvents—a concept summarized by the rule *"like dissolves like"*, which is essential for understanding drug solubility and delivery.

Matter exists in four states—solid, liquid, gas, and plasma—all of which have roles in healthcare, from mineralized bone tissue (solids) to sterilization tools using plasma. Importantly, many ionic compounds dissociate into electrolytes (such as Na^+, K^+, Ca^{2+}, Cl^-), which are essential for nerve signaling, muscle contraction, cardiac rhythm, and maintaining fluid balance.

Nurses apply this foundational chemistry daily: interpreting diagnostic tests involving isotopes, administering medications based on solubility and polarity, correcting electrolyte imbalances, and operating plasma-based medical equipment. All of these clinical practices are rooted in the atomic and molecular principles discussed in this chapter.

EXERCISES

Multiple Choice Questions

1. Which subatomic particle carries a positive electrical charge?
   a) Electron
   b) Neutron
   c) Proton
   d) Ion
2. What characterizes the formation of an ionic bond?
   a) Sharing of electrons
   b) Transfer of electrons
   c) Formation of isotopes
   d) Hydrogen bonding
3. Which state of matter lacks both a definite shape and a definite volume?
   a) Solid
   b) Liquid
   c) Gas
   d) Plasma
4. Water is best classified as a:
   a) Nonpolar solvent
   b) Polar solvent
   c) Metallic compound
   d) Ionic salt
5. Which substance dissociates into Na^+ and Cl^- ions in aqueous solution?
   a) Glucose
   b) Water
   c) Sodium chloride
   d) Ethanol
6. Which type of radiation has the greatest ability to penetrate tissue and is widely used in both imaging and cancer treatment?
   a) Alpha (α) radiation
   b) Beta (β) radiation
   c) Gamma (γ) rays
   d) Positrons (β^+)

7. Which medical imaging device is designed to detect gamma rays emitted by isotopes like technetium-99m?
   a) PET scanner
   b) CT scanner
   c) Gamma camera
   d) X-ray tube
8. Iodine-131 is most commonly used in the medical field to:
   a) Image bone structure
   b) Treat thyroid disorders
   c) Diagnose brain function
   d) Assess lung ventilation
9. In PET imaging, which type of radiation is emitted and interacts with electrons to produce gamma rays?
   a) Alpha particles
   b) Beta minus particles
   c) Gamma rays
   d) Positrons (β^+)

True or False

1. Electrons carry a positive electrical charge.
2. Isotopes of an element have the same number of protons but different numbers of neutrons.
3. Covalent bonds are formed when atoms gain or lose electrons.
4. Polar molecules dissolve well in water.
5. Plasma is the most common state of matter in the human body.
6. Alpha particles can deeply penetrate tissue and are suitable for external imaging.
7. Gamma rays and X-rays are both types of electromagnetic radiation.
8. PET scans track metabolic activity using isotopes that emit positrons.
9. Beta radiation is completely harmless and requires no protective measures.

Fill in the Blanks

1. The ____________ of an atom is defined by the number of protons in its nucleus.
2. The ____________ of an element is the total number of protons and neutrons.
3. ____________ bonds form when two atoms share electron pairs.
4. The principle of ____________ explains why polar substances dissolve in polar solvents.
5. In clinical practice, ____________ like Na^+ and K^+ are essential for nerve and muscle function.
6. ____________ radiation, composed of two protons and two neutrons, has low penetration and is used for internal therapy.
7. Gamma radiation is commonly detected using a ____________ in nuclear medicine.
8. ____________-131 is a radioactive isotope used in treating thyroid conditions.
9. A ____________ is a device that produces high-energy X-rays for external cancer treatment.

Short Answer Questions

1. What is the fundamental difference between an element and a compound?
2. Define isotopes and provide a clinical example of their use.
3. Explain why ionic compounds such as sodium chloride dissolve easily in water.
4. How does molecular polarity influence drug absorption and distribution in the human body?

Critical Thinking

1. A patient is receiving IV fluids containing electrolytes. How might an excess or deficiency in potassium ions (K^+) affect nerve conduction and muscle activity?
2. Plasma is often described as the "fourth state of matter." In which medical technologies or treatments might plasma be encountered?

3. Some medications are nonpolar and poorly soluble in water. As a nurse, what considerations should be taken into account when administering such drugs?
4. A new drug forms strong covalent bonds with its carrier molecule. Discuss one possible benefit and one potential drawback of this interaction in the body.
5. Why is it essential for nurses to distinguish between alpha, beta, and gamma radiation in clinical environments?
6. A patient is scheduled for a PET scan using fluorine-18. What safety guidelines should the nurse communicate before and after the procedure?
7. Compare the use of gamma rays in diagnostic imaging versus therapeutic radiation. How do the objectives and precautions differ in each application?

Chapter 4

Organic Molecules and Body Functions

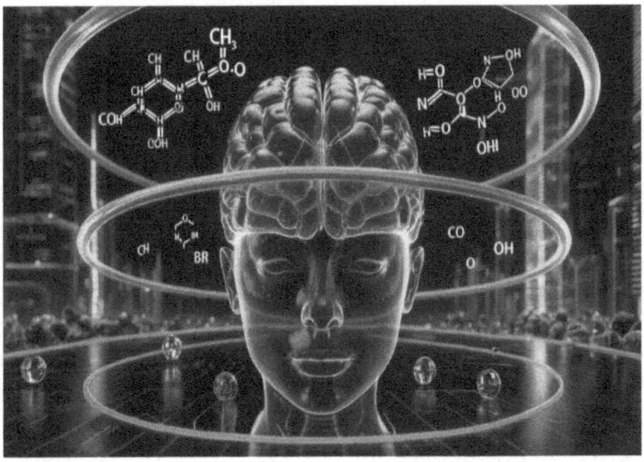

Chapter 4: Organic Molecules and Body Functions

Learning Objectives

By the end of this chapter, you should be able to:

1. Identify key types of organic compounds in the body.
2. Recognize major functional groups and their roles.
3. Classify organic compounds and understand their biological use.
4. Distinguish between saturated and unsaturated fats.
5. Understand basic IUPAC naming rules.
6. Explain isomerism and its medical importance

Organic molecules form the chemical foundation of all living systems. For healthcare professionals, understanding these compounds is essential to interpreting physiological processes, recognizing disease mechanisms, and understanding how drugs work. This chapter introduces the major categories of organic compounds—including hydrocarbons, alcohols, aldehydes, ketones, and carboxylic acids—through the lens of functional groups, which are specific atomic arrangements that determine molecular behavior.

Key functional groups such as hydroxyl (–OH), carbonyl (C=O), carboxyl (–COOH), and amino ($-NH_2$) influence solubility, reactivity, and biological activity. For example, ethanol (an alcohol) is widely used as a disinfectant, while ketones build up in diabetic ketoacidosis. Many medications, such as aspirin, rely on specific functional groups for their therapeutic effects.

By studying these groups, students will understand how structure relates to function in metabolism, pharmacology, and clinical diagnostics. The chapter also highlights the biological roles of hydrocarbons in energy storage and the clinical relevance of alcohols, acids, and ketones in health and disease.

Organic Compounds and Functional Groups

Organic compounds are molecules primarily composed of carbon and hydrogen, often with additional atoms such as oxygen, nitrogen, sulfur, or phosphorus. In biological and medical contexts, these compounds form the chemical basis for all macromolecules essential to life—including carbohydrates, proteins, lipids, and nucleic acids.

What gives organic compounds their chemical diversity and biological specificity is the presence of **functional groups**—specific groupings of atoms that determine a molecule's chemical properties and reactivity. These groups participate in chemical reactions, influence molecular shape, and enable interactions with enzymes, receptors, and other biomolecules in the body.

Key Functional Groups and Their Roles:

- **Hydroxyl group (-OH):** Found in alcohols and sugars such as glucose, the hydroxyl group increases a molecule's polarity and solubility in water, making it essential for cellular energy and metabolism.
- **Carbonyl group (C=O):** This group is present in both aldehydes and ketones and is involved in many metabolic reactions. For example, ketone bodies (containing carbonyls) are produced during fasting or in uncontrolled diabetes.
- **Carboxyl group (-COOH):** Found in amino acids and fatty acids, the carboxyl group acts as a weak acid and is important in maintaining pH balance and forming peptide bonds in proteins.
- **Amino group ($-NH_2$):** Present in amino acids and many drug molecules, the amino group acts as a base and contributes to the structure and charge of biomolecules.

Biological Examples:

- **Glucose** contains multiple hydroxyl groups, allowing it to dissolve in blood and be readily transported and metabolized by cells.
- **Amino acids**, the building blocks of proteins, contain both amino and carboxyl groups, enabling them to link together into peptides and proteins that perform countless biological functions.

Understanding these functional groups is crucial for health professionals, as they underpin the chemical behavior of nutrients, drugs, and metabolic byproducts. In clinical settings, knowledge of functional groups helps explain how drugs bind to receptors, how enzymes catalyze reactions, and how lab results (e.g., presence of ketones or blood urea nitrogen) relate to molecular processes in the body.

Notes for Nursing and Allied Health:

- Changes in **ketone levels** (carbonyl group) are a key diagnostic indicator in **diabetic ketoacidosis**.
- **Carboxyl and amino groups** are essential for **protein folding and buffering blood pH**.
- Many **drug molecules** mimic these functional groups to **bind to enzymes or receptors** effectively.

Table 4.1 Major Organic Functional Groups and Medical Examples

Functional Group	Structure	Found In	Biological / Medical Example
Hydroxyl	$–OH$	Alcohols, sugars	Glucose (energy source); Ethanol (used as antiseptic)
Carbonyl	$>C{=}O$	Aldehydes, ketones	Acetone (ketone body in diabetes); Formaldehyde (preservative)
Carboxyl	$–COOH$	Carboxylic acids	Fatty acids; Amino acids (form peptide bonds in proteins)
Amino	$–NH_2$	Amines, amino acids	Glycine (basic amino acid); protein structure
Methyl	$–CH_3$	Hydrocarbons	Lipids; methylation of DNA (epigenetic regulation)
Phosphate	$–PO_4^{2-}$	Organic phosphates	ATP (energy transfer); backbone of DNA/RNA
Sulfhydryl	$–SH$	Thiols	Cysteine (forms disulfide bonds); protein stabilization

Hydrocarbons

Hydrocarbons are the simplest organic compounds, composed solely of carbon and hydrogen atoms. They form the structural backbone of many complex biomolecules and serve as a fundamental starting point in understanding organic chemistry. Although not highly reactive on their own, hydrocarbons become biologically significant when functional groups are attached to their carbon skeletons.

Types of Hydrocarbons:

Alkanes (Single Bonds Only):

Alkanes are the simplest type of hydrocarbons, characterized by single bonds between carbon atoms. They follow the general molecular formula C_nH_{2n+2} and are classified as saturated hydrocarbons because each carbon atom is bonded to the maximum possible number of hydrogen atoms. This structural saturation makes alkanes relatively stable and less reactive compared to other hydrocarbon groups. A common example is methane (CH_4), the simplest alkane, which exists as a natural gas and is also produced as a byproduct during digestive processes in the human gut.

Alkenes (One or More Double Bonds):

Alkenes are a class of unsaturated hydrocarbons that contain at least one carbon–carbon double bond, giving them the general formula C_nH_{2n}. The presence of double bonds makes alkenes more chemically reactive than alkanes and allows for various types of chemical reactions important in biological and industrial processes. A well-known example is ethene (C_2H_4), which plays a key role in plant hormone signaling, particularly in the ripening of fruit, and also serves as a foundational compound in the synthesis of various pharmaceutical drug precursors.

Alkynes (One or More Triple Bonds):

Alkynes are a group of unsaturated hydrocarbons distinguished by the presence of one or more carbon–carbon triple bonds, and they follow the general formula C_nH_{2n-2}. The triple bond introduces significant reactivity, making alkynes more chemically active than both alkanes and alkenes. A notable example is ethyne, also known as acetylene (C_2H_2), which is widely used in medical sterilization technologies due to its high energy and reactivity, enabling effective disinfection of heat-sensitive medical instruments.

Aromatic hydrocarbons:

Aromatic hydrocarbons are a distinct class of organic compounds that contain one or more planar ring structures with delocalized π-electrons, typically based on the benzene ring (C_6H_6). These compounds are known for their stability, due to resonance, and their unique chemical behavior compared to other hydrocarbons. Aromatic hydrocarbons are often lipophilic and can easily penetrate biological membranes, which contributes to their widespread use in pharmaceuticals. A classic example is benzene, which serves as the structural core for many drugs, dyes, and industrial chemicals. Despite their usefulness, some aromatic compounds can be toxic or carcinogenic, so their handling in medical and laboratory settings requires caution.

Saturated vs. Unsaturated hydrocarbons :

In healthcare and nutrition, the distinction between saturated and unsaturated hydrocarbons is essential for understanding lipid chemistry:

Saturated hydrocarbons:

Saturated hydrocarbons are organic compounds composed entirely of carbon and hydrogen atoms connected by single bonds. Each carbon atom in these molecules is bonded to the maximum possible number of hydrogen atoms, which is why they are referred to as "saturated." Structurally stable and less reactive, saturated hydrocarbons serve as the foundation for substances like alkanes. In biological contexts, saturated hydrocarbons are found in the fatty acid chains of saturated fats, which are solid at room temperature—such as butter and animal fats. Although they are an important energy source, excessive intake is linked to increased levels of LDL cholesterol and a higher risk of cardiovascular disease.

Unsaturated hydrocarbons:

Unsaturated hydrocarbons are organic compounds that contain at least one double or triple bond between carbon atoms in their molecular structure. This category includes alkenes (with one or more double bonds) and alkynes (with one or more triple bonds). These bonds introduce bends or kinks in the hydrocarbon chain, affecting the compound's physical and chemical properties. In biological systems, unsaturated hydrocarbons are typically found in unsaturated fatty acids, which are usually liquid at room temperature—such as olive oil and other plant-based oils. These fats are associated with health benefits, including improved cardiovascular function and reduced LDL cholesterol levels, making them a preferred dietary choice over saturated fats.

Clinical Relevance:

Hydrocarbon-based lipids play vital roles in human physiology and medicine. Phospholipids, which contain hydrocarbon tails, are essential structural components of cell membranes, providing fluidity and selective permeability. Additionally, lipids serve as long-term energy storage molecules, especially in adipose tissue. In clinical practice, several anesthetics and intravenous medications are synthesized from hydrocarbon derivatives due to their ability to interact with lipid membranes and biological targets. Importantly, the degree of saturation in dietary fats significantly impacts health: saturated fats tend to raise LDL cholesterol and promote inflammation, whereas unsaturated fats can improve lipid profiles and reduce cardiovascular risk. Understanding these distinctions is crucial for guiding dietary recommendations and managing conditions like heart disease and metabolic syndrome.

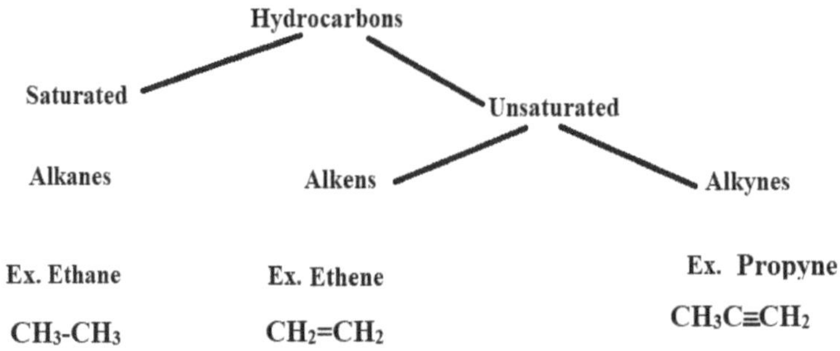

Figure 4. 1. saturated vs. unsaturated hydrocarbons.

"Industrial Oil Refinery Plant in Desert Environment"

The following table outlines the main types of hydrocarbons based on their bond structure, with examples and clinical relevance. Understanding these differences helps explain their roles in biological systems and pharmaceutical applications.

Table 4.2 Types of Hydrocarbons and Their Medical Relevance

Type	Formula	Key Property	Example	Medical/ Biological Use
Alkanes	C_nH_{2n+2}	Saturated, low reactivity	Methane (CH_4)	Digestive gas, anesthetic derivatives
Alkenes	C_nH_{2n}	Unsaturated, more reactive	Ethene (C_2H_4)	Hormone signaling, drug synthesis
Alkynes	C_nH_{2n-2}	Highly reactive, linear shape	Ethyne (C_2H_2)	Sterilization gases, pharma precursors
Aromatics	—	Stable, lipophilic	Benzene	Basis of many drugs (aspirin, paracetamol)

Carbon Chain Prefixes in Organic Nomenclature

In IUPAC nomenclature, the name of an organic compound is based on the **number of carbon atoms** in its longest continuous chain. Each number corresponds to a specific **prefix**, which serves as the root of the compound's name. This system is essential for naming hydrocarbons (alkanes, alkenes, alkynes) and other organic compounds such as alcohols, acids, and ketones.

Table 4.3 Carbon Chain Prefixes in Organic Chemistry (IUPAC)

Number of Carbon Atoms	Prefix	Example (Alkane)	Formula
1	**Meth-**	Methane	CH_4
2	**Eth-**	Ethane	C_2H_6
3	**Prop-**	Propane	C_3H_8
4	**But-**	Butane	C_4H_{10}
5	**Pent-**	Pentane	C_5H_{12}
6	**Hex-**	Hexane	C_6H_{14}
7	**Hept-**	Heptane	C_7H_{16}
8	**Oct-**	Octane	C_8H_{18}
9	**Non-**	Nonane	C_9H_{20}
10	**Dec-**	Decane	$C_{10}H_{22}$

Key Notes: These prefixes are combined with suffixes like:

- **–ane** (alkanes, single bonds)
- **–ene** (alkenes, double bonds)
- **–yne** (alkynes, triple bonds)

- Example:

  - **Ethane** = 2 carbon atoms + a single bond
  - **Propene** = 3 carbon atoms + a double bond
  - **Butyne** = 4 carbon atoms + a triple bond

Understanding this naming system helps health science students recognize and interpret the structure of drug molecules, metabolites, and biochemical compounds in clinical settings.

Exercise : What are the correct names of the organic compounds to fill in the blanks?

1. 1 carbon atom + a single bond = _______

2. 2 carbon atoms + a double bond = _______

3. 3 carbon atoms + a triple bond = _______

4. 5 carbon atoms + a single bond = _______

5. 6 carbon atoms + a double bond = _______

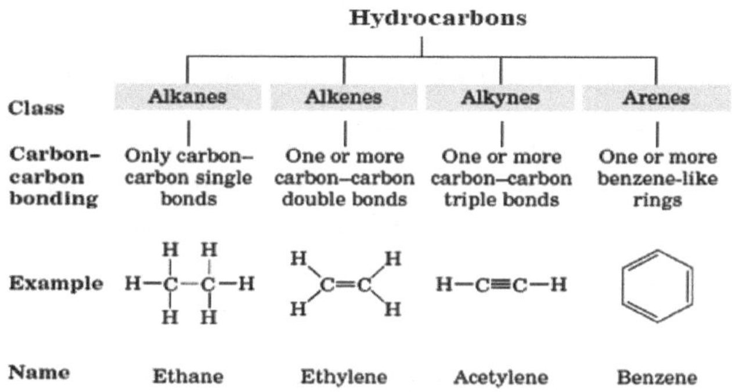

Figure 4. 2. Classification of hydrocarbons

Isomers: Same Formula, Different Structures

In organic chemistry, **isomers** are compounds that have the **same molecular formula** but **different structural arrangements** of atoms. This difference in structure leads to different physical and chemical properties—even though the molecules are made from the same types and numbers of atoms.

Understanding isomers is crucial in medicine and pharmacology, as the **biological activity of drugs** often depends on their exact 3D structure.

The Table 4.4 below summarizes the main types of isomers, with examples and clinical relevance—highlighting how small structural differences can lead to major biological effects.

Table 4.4 Types of Isomers

Type	Definition	Example	Clinical Relevance
Structural isomers	Same formula, different atom connectivity	Butane vs. Isobutane (C_4H_{10})	Different boiling points and reactivity
Geometric (cis-trans)	Same connectivity, different spatial arrangement	Cis- vs. Trans-fats	Trans fats raise cardiovascular risk
Enantiomers (optical)	Non-superimposable mirror images with chirality	L-Dopa vs. D-Dopa	L-Dopa treats Parkinson's; D-Dopa inactive
Diastereomers	Stereoisomers not mirror images	Glucose vs. Galactose	Different metabolic pathways

Why Isomers Matter in Health Professions:

Isomers are critically important in healthcare because the **effectiveness, safety, and metabolism** of many drugs depend on the **three-dimensional arrangement** of their molecules.

A powerful example is **thalidomide**: one isomer treated **morning sickness** in pregnant women, while its **mirror-image** caused **birth defects**—an event that changed drug regulations worldwide. This underscores the necessity of understanding isomerism in **drug design**.

Moreover, human **enzymes and receptors are stereospecific**—they interact only with molecules that have the correct spatial shape. For example, **D-glucose** is readily metabolized by human cells, while **L-glucose** is not, despite having the same molecular formula.

These differences are vital in **pharmacology, diagnostics, and patient safety**, making isomerism a core concept for healthcare professionals.

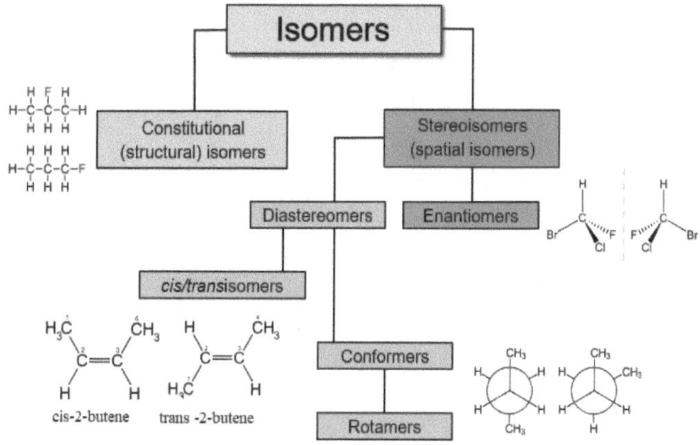

Figure 4.3. Classification of Isomers

The Importance of Chirality and Isomerism in Healthcare and Drug Design

Chirality is a molecular property where a compound cannot be superimposed on its **mirror image**, much like your right and left hands. This occurs when a molecule has a **chiral center**—typically a carbon atom bonded to **four different atoms or groups**.

Such molecules exist in two forms called **enantiomers**, which are **mirror images** of each other but not identical. Their biological effects can differ dramatically.

To distinguish between them, chemists use the **R (rectus)** and **S (sinister)** configuration system, based on a set of **priority rules** that consider the spatial orientation of attached groups. A mixture containing equal amounts of both forms is called a **racemic mixture**, which is usually **less effective biologically**, since **only one enantiomer** often fits the body's enzymes or receptors.

Understanding chirality, chiral centers, and R/S naming is **essential in pharmacology**, because one enantiomer may be **therapeutic** while the other could be **inactive or harmful**.

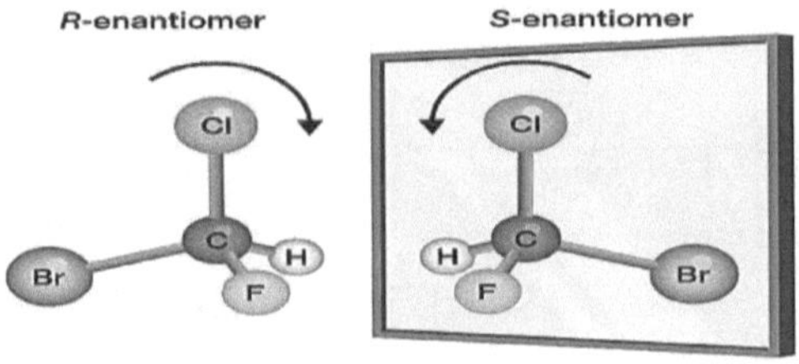

Figure 4.4. Chirality :The R- and S-enantiomers

D and L System in Biological Molecules

In addition to the R/S system, another important method for classifying enantiomers—especially in sugars and amino acids—is the D/L system. This naming convention is based on the molecule's structural relationship to glyceraldehyde, the simplest chiral carbohydrate. In this system, a molecule is labeled D (dextro) if the hydroxyl or amino group is positioned on the right in a Fischer projection, and L (levo) if it is on the left. Importantly, D and L do not refer to the direction in which the compound rotates plane-polarized light, nor do they correspond directly to R and S designations. In biological systems, most naturally occurring amino acids are in the L-form, while sugars (such as glucose) are typically in the D-form. This specificity is critical because enzymes and receptors in the body are highly selective and often interact with only one of the two forms, making the D/L distinction essential in biochemistry, nutrition, and drug development.

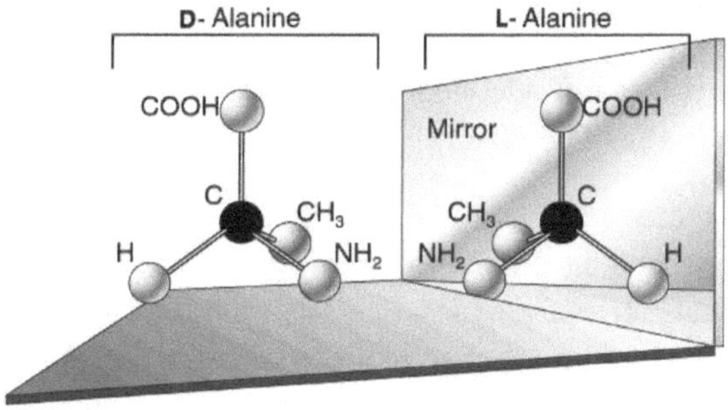

Figure 4.5. The D- and L-enantiomers of Alanine

Alcohols, Aldehydes, Ketones, and Carboxylic Acids

Organic compounds containing oxygen-based functional groups play essential roles in both biological systems and medical applications. Among the most clinically relevant groups are **alcohols**, **aldehydes**, **ketones**, and **carboxylic acids**. These molecules differ in structure, reactivity, and physiological function—but all are vital to health and pharmacology.

Alcohols (–OH Group)

Alcohols are organic compounds that contain one or more hydroxyl (**–OH**) groups attached to a saturated carbon atom. They are **polar**, capable of forming **hydrogen bonds**, and are highly relevant in both biological systems and medical applications.

Structure and Naming of Alcohols

The general formula of alcohols is R–OH, where R represents an alkyl group derived from a hydrocarbon chain. In IUPAC nomenclature, alcohols are named by identifying the longest carbon chain containing the –OH group and replacing the "–e" ending of the corresponding alkane with "–ol." For example, methane becomes methanol (CH_3OH), and ethane becomes ethanol (CH_3CH_2OH). The position of the –OH group is indicated by numbering the carbon chain to give the hydroxyl group the lowest possible number when necessary. This systematic naming helps clearly identify the structure and function of alcohols in both chemical and biomedical contexts.

Classification of Alcohols

1. Classification of Alcohols Based on the Type of Carbon Atom Bonded to the –OH Group

Alcohols can be classified structurally as primary, secondary, or tertiary depending on the number of carbon atoms bonded to the carbon that carries the hydroxyl group (–OH). A primary (1°) alcohol has the –OH group attached to a carbon atom that is bonded to only one other carbon; a common example is ethanol, which is widely used in beverages and as a disinfectant. A secondary (2°) alcohol has the hydroxyl-bearing carbon bonded to two other carbon atoms, as seen in isopropanol (commonly known as rubbing alcohol), which is used as an antiseptic and solvent. In contrast, a tertiary (3°) alcohol has the –OH group attached to a carbon atom that is connected to three other carbon atoms. An example is tert-butyl alcohol, which is used in synthetic chemistry and fuel additives. This classification is important because it influences the alcohol's reactivity, especially in oxidation reactions and biochemical pathways.

Table 4.5 Classification of Alcohols Based on the Type of Carbon Atom Bonded to the –OH Group

Type	Definition	Example
Primary (1°)	–OH attached to a carbon bonded to one other carbon	Ethanol
Secondary (2°)	–OH attached to a carbon bonded to two other carbons	Isopropanol
Tertiary (3°)	–OH attached to a carbon bonded to three carbons	Tert-butyl alcohol

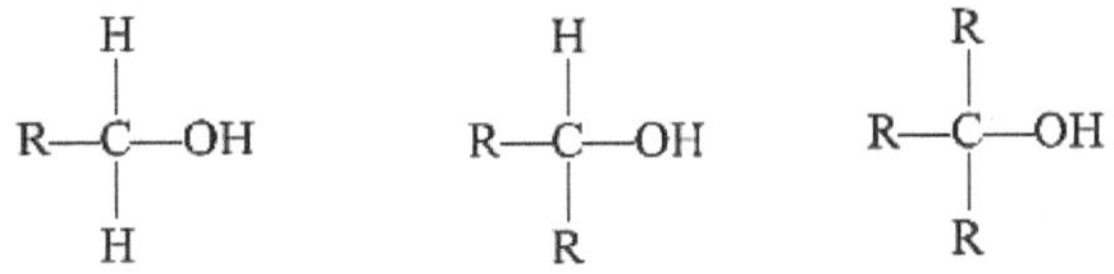

primary alcohol **secondary alcohol** **tertiary alcohol**

Figure 4.6 Classification of Alcohols Based on the Type of Carbon Atom Bonded to the –OH Group

2. Classification Based on the Number of Hydroxyl Groups (-OH)

Alcohols can also be classified according to the number of hydroxyl (–OH) groups present in their molecular structure. Monohydric alcohols contain only one –OH group; a common example is ethanol (CH_3CH_2OH), which is used in alcoholic beverages and medical applications. Dihydric alcohols, also called glycols, have two –OH groups, such as ethylene glycol ($HO–CH_2–CH_2–OH$), which is used in antifreeze and coolant solutions. Trihydric alcohols possess three hydroxyl groups, with glycerol ($HO–CH_2–CHOH–CH_2OH$) being a well-known example used in pharmaceuticals, cosmetics, and food products. Lastly, polyhydric alcohols have more than three –OH groups, like sorbitol and mannitol, which are commonly used as sugar substitutes and in medical formulations. This classification highlights the diversity of alcohols in both chemical behavior and practical applications.

Table 4.6 Types of Alcohols Based on –OH Groups

Type	# of –OH Groups	Example	Medical/Biological Use
Monohydric	1	Ethanol	Antiseptic, CNS depressant
Dihydric	2	Ethylene glycol	Industrial use (toxic); precursor in biosystems
Trihydric	3	Glycerol	Moisturizers, laxatives, metabolic intermediate

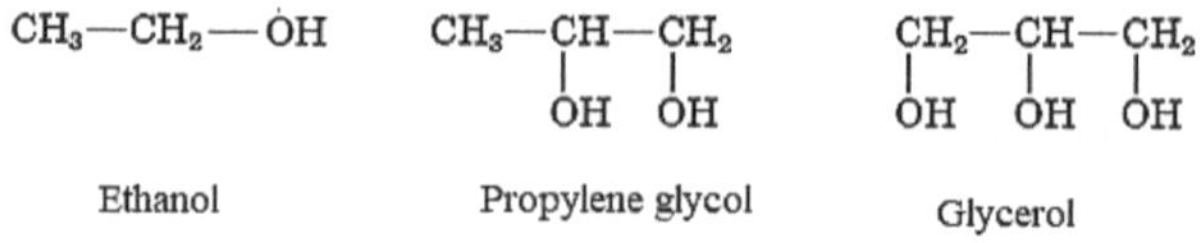

Figure 4.7 Types of Alcohols Based on –OH Groups

Biological and Medical Importance of Alcohols

Alcohols play significant roles in biological systems, medicine, and clinical settings. **Ethanol** is one of the most widely used alcohols, serving as a disinfectant, antiseptic, and solvent in pharmaceutical preparations. It also acts as a central nervous system (CNS) depressant and is metabolized in the liver to acetaldehyde, which is associated with various physiological effects.

Another important alcohol is **glycerol (also known as glycerin)**, a trihydroxy alcohol containing three hydroxyl groups. Glycerol is a natural component of triglycerides and phospholipids and is used in medical creams, as a moisturizer, and as an osmotic laxative.

Cholesterol, although structurally different, is classified as a **steroid alcohol (sterol)**. It is vital for maintaining cell membrane stability and serves as a precursor for the synthesis of steroid hormones and vitamin D.

In pharmaceutical applications, alcohols are commonly used as solvents due to their ability to dissolve both polar and nonpolar compounds. The presence of hydroxyl groups in many drugs influences their solubility, metabolism, and ability to bind to biological targets. For instance, **salicylic acid**, a precursor to aspirin, contains a hydroxyl group that is key to its biological activity.

In laboratory and clinical environments, **ethanol and isopropanol** are extensively used as skin antiseptics and surface disinfectants due to their antimicrobial properties. Additionally, **glycerol** plays a critical role in the **cryopreservation of cells and tissues**, helping protect biological samples during freezing and thawing processes.

These diverse roles highlight the essential nature of alcohols in both health and science.

Aldehydes and Ketones

Aldehydes and ketones are important classes of organic compounds characterized by the presence of a carbonyl group (C=O). The primary difference between them lies in the location of this functional group within the carbon chain. In **aldehydes**, the carbonyl group is located at the end of the chain, where the carbon atom of the C=O bond is attached to at least one hydrogen atom. This terminal positioning contributes to their generally higher reactivity compared to ketones. A common example is **formaldehyde (HCHO)**, which is widely used in tissue preservation and as a disinfectant.

In contrast, **ketones** have the carbonyl group positioned within the carbon chain, flanked by two other carbon atoms. This internal placement influences their chemical properties, often making them less reactive but still biologically significant. A well-known example is **acetone (CH_3COCH_3)**, a versatile solvent and a naturally occurring metabolic compound.

The structural difference in carbonyl positioning impacts not only their **reactivity and solubility**, but also their **biological roles**, such as in gluconeogenesis (the formation of glucose from non-carbohydrate sources) and the production of **ketone bodies**, which serve as alternative energy sources during prolonged fasting or in uncontrolled diabetes.

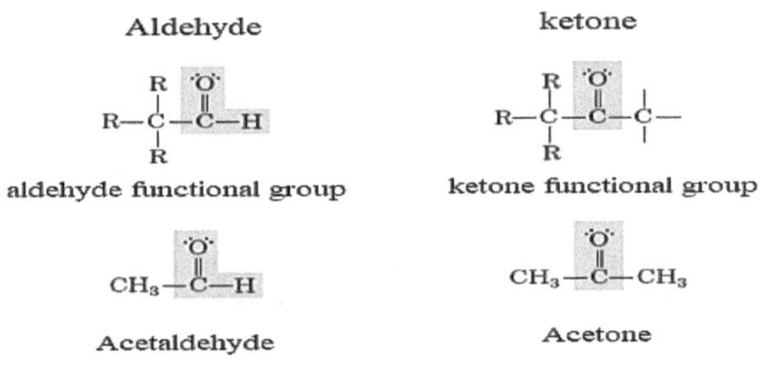

Figure 4.8 Aldehydes and Ketones –CO Groups

Aldehydes

Definition and Structure

Aldehydes are a class of organic compounds defined by the presence of a carbonyl group (C=O) bonded to at least one hydrogen atom. This carbonyl group is always located at the **end** of a carbon chain, which distinguishes aldehydes from other carbonyl-containing compounds like ketones. The **general formula** for aldehydes is **R–CHO**, where *R* represents a hydrogen atom or a hydrocarbon group (alkyl or aryl). The functional group –CHO consists of a carbon atom double-bonded to an oxygen atom and single-bonded to a hydrogen atom.

This structure imparts **high reactivity** to aldehydes, particularly in **oxidation reactions** and **nucleophilic addition**. Two well-known examples are **formaldehyde (HCHO)**, used in disinfection and tissue preservation, and **acetaldehyde (CH_3CHO)**, an important intermediate in metabolism and the chemical industry.

IUPAC Naming Rules for Aldehydes

In the IUPAC system, aldehydes are named using specific rules that reflect the presence of the carbonyl functional group. The suffix **"-al"** is used to indicate an aldehyde. To name an aldehyde, begin with the name of the **parent alkane** corresponding to the number of carbon atoms, and then replace the terminal "**-e**" with "**-al**."

For example:

- **Methane** becomes **methanal** (HCHO)
- **Ethane** becomes **ethanal** (CH_3CHO)

The **carbon atom of the aldehyde group (–CHO)** is always assigned **position 1**, so its location does not need to be explicitly numbered in straight-chain aldehydes. When other substituents are present, the carbon chain is numbered starting from the aldehyde carbon to ensure **correct positional identification** of the substituents.

Biological and Clinical Significance of Aldehydes

Aldehydes are highly reactive organic compounds that play vital roles in biological systems and clinical applications. Their structural simplicity and the presence of a reactive carbonyl group make them key intermediates in metabolic pathways and valuable tools in medical diagnostics.

In the human body, certain aldehydes have essential physiological functions. For instance, **acetaldehyde (ethanal)** is formed during the hepatic metabolism of **ethanol** and is responsible for many of the negative effects of alcohol consumption, including **hangovers** and **liver toxicity**. Clinically, elevated levels of acetaldehyde can cause a **sharp, pungent breath odor**, often described as resembling **fermented fruit or rotten apples**, which serves as a diagnostic clue in cases of **alcohol intoxication** or **hepatic dysfunction**.

Another biologically important aldehyde is **retinal**, a derivative of **vitamin A**, which plays a central role in the **visual cycle**. Retinal binds to the protein **opsin** to form **rhodopsin**, the light-sensitive pigment in the rod cells of the retina, which is essential for **vision in low-light conditions**.

Aldehydes also play a significant role in **clinical diagnostics**. For example, the aldehyde group in **glucose** (in its open-chain form) enables it to function as a **reducing sugar**—a property utilized in traditional chemical tests for **diabetes mellitus**. In **Fehling's test**, glucose reduces copper(II) ions to red **copper(I) oxide**, while in **Tollens' test**, it reduces silver ions to metallic silver, forming a **distinctive silver mirror**. These reactions are used to detect **glucose in urine (glycosuria)**, commonly seen in uncontrolled diabetes.

In healthcare settings, aldehydes have widespread applications. **Formaldehyde (methanal)** is widely used as a **preservative for biological specimens** and as a **disinfectant for surgical equipment**. However, due to their high reactivity and potential toxicity, many aldehydes must be handled with strict **safety precautions**.

In summary, a thorough understanding of the **chemical behavior**, **diagnostic utility**, and **biological relevance** of aldehydes is essential for healthcare professionals. This knowledge supports effective patient evaluation and promotes the safe use of aldehyde-containing substances in clinical environments.

Table 4.7 Biologically and Clinically Significant Aldehydes

Compound	**Chemical Formula**	**Biological / Clinical Role**
Formaldehyde (Methanal)	HCHO	Colorless, highly reactive gas with a pungent odor. Used in histology and tissue preservation. Acts as a potent disinfectant but is toxic and requires strict handling.
Acetaldehyde (Ethanal)	CH_3CHO	Volatile intermediate in ethanol metabolism; contributes to hangover symptoms and liver toxicity. Breath may carry a fruity, aldehydic odor in alcohol-related cases.
Retinal (Vitamin A Aldehyde)	$C_{20}H_{28}O$	Essential for vision. Derived from vitamin A and combines with opsin to form rhodopsin, enabling light detection in the retina, particularly in low-light conditions.
Glucose (Open-Chain Form)	$C_6H_{12}O_6$	Reducing sugar with an aldehyde group. Central to energy metabolism and diagnostic testing for diabetes (e.g., Fehling's and Tollens' tests to detect glycosuria).

Ketones

Definition and Structure

Ketones are a class of organic compounds characterized by the presence of a **carbonyl group (C=O)** bonded to **two carbon atoms**. Unlike aldehydes, the carbonyl group in ketones is **never located at the end** of the carbon chain but is always situated **within** it. This internal placement is a key distinguishing feature.

The **general formula** for ketones is **R–CO–R′**, where **R** and **R′** represent alkyl or aryl groups, which may be identical or different. In ketones, the carbonyl carbon is **double-bonded to an oxygen atom** and **single-bonded to two carbon atoms**, giving the molecule a **planar and polar** structure. This polarity contributes to their **moderate reactivity**, especially in **nucleophilic addition reactions**.

Common ketones include:

- **Acetone (CH_3COCH_3)** – a widely used solvent in industry and pharmaceuticals.
- **Butanone ($CH_3COCH_2CH_3$)** – used in industrial processes and laboratory settings.

In biological systems, ketones also have significant metabolic importance. They are involved in the production of **ketone bodies**, such as **acetoacetate** and **β-hydroxybutyrate**, which serve as alternative energy sources during **prolonged fasting**, **starvation**, or **uncontrolled diabetes mellitus**.

IUPAC Naming Rules of Ketones

In the IUPAC system, ketones are named by identifying the **longest continuous carbon chain** that contains the **carbonyl group (C=O)** and replacing the **"-e"** ending of the parent alkane with **"-one."** The carbon chain is **numbered** in a way that assigns the **lowest possible number** to the carbonyl carbon, and this number is included in the name to indicate the **position** of the carbonyl group.

For example: A four-carbon ketone with the carbonyl group on the second carbon is named **butan-2-one**.

When **substituent groups** are present, they are:

1. **Named and numbered** according to their position on the main chain.

2. Listed in **alphabetical order** before the base name.

Example: 3-methylbutan-2-one refers to a ketone with four carbon atoms, a **methyl group on carbon 3**, and a **carbonyl group on carbon 2**.

In **simple symmetrical ketones** like **propanone (acetone)**, where the position of the carbonyl group is unambiguous, the number is often **omitted**.

This systematic naming approach ensures **clarity** in identifying ketone structures, which is especially useful in **clinical**, **pharmaceutical**, and **laboratory** settings.

Biological and Clinical Significance of Ketones

Ketones play a vital role in human metabolism, particularly during periods when glucose availability is limited, such as fasting, prolonged physical exertion, or uncontrolled diabetes mellitus. In these states, the liver converts fatty acids into ketone bodies—namely acetoacetate, β-hydroxybutyrate, and acetone—which act as alternative energy sources, especially for energy-demanding organs such as the brain, heart, and skeletal muscles. This metabolic adaptation helps sustain energy production when glucose supplies are insufficient.

From a clinical perspective, the presence of ketones in blood or urine is a critical diagnostic marker. In conditions like diabetic ketoacidosis (DKA)—a life-threatening complication of uncontrolled type 1 diabetes—the overproduction of ketone bodies leads to metabolic acidosis, electrolyte imbalances, and severe dehydration, requiring urgent medical intervention. Measurement of ketone levels is also useful in monitoring ketogenic or low-carbohydrate diets, and in evaluating metabolic status during starvation, illness, or fasting states.

Beyond their metabolic role, certain steroid hormones (e.g., testosterone, progesterone) and various pharmaceutical agents contain ketone functional groups, which contribute to their biological activity. The carbonyl group in ketones enhances their chemical reactivity, enabling interactions with enzymes and receptors in biological systems.

Thus, a solid understanding of ketones is essential for assessing metabolic function, managing diabetes, interpreting biochemical test results, and appreciating their role in drug action and hormone physiology.

Table 4.8: Biologically and Clinically Significant of Ketones

Ketone Name	Formula	Key Role
Acetone	CH_3COCH_3	Ketone body from fasting/diabetes; excreted
Acetoacetate	$CH_3COCH_2COO^-$	Liver-produced ketone; energy source
β-Hydroxybutyrate	$HOCH_2CH_2COO^-$	Major ketone in ketosis; energy source
Propanone	CH_3COCH_3	Industrial acetone; detected in diabetic breath
Testosterone	$C_{19}H_{28}O_2$	Steroid hormone for male development
Progesterone	$C_{21}H_{30}O_2$	Steroid hormone regulating menstrual cycle
Cortisone	$C_{21}H_{28}O_5$	Anti-inflammatory corticosteroid

Laboratory Detection of Aldehydes and Ketones

Aldehydes and ketones are commonly detected and differentiated in the laboratory using classical chemical tests that rely on their **reactivity with specific reagents**. These tests take advantage of the fact that **aldehydes are more easily oxidized** than ketones, while both contain the **carbonyl group (C=O)**. Such tests are valuable in clinical biochemistry, organic synthesis, and compound identification.

Table 4.9: Laboratory Detection of Aldehydes and Ketones

Test Name	Purpose	Reagent	Positive Result	Ald	Ket
2,4-DNP Test	Detects carbonyl groups (C=O) in both classes	2,4Dinitro-phenylhydrazine	Yellow, orange, or red precipitate	□	□
Tollen's Test	Specific for aldehydes	Ammoniacal silver nitrate (Ag^+)	Shiny silver mirror	□	□
Fehling's Test	Detects reducing aldehydes	Fehling's solution (Cu^{2+} complex)	Brick-red/orange precipitate (Cu_2O)	□	□
Benedict's Test	Detects reducing sugars (acts like aldehydes)	Benedict's Solution (Cu^{2+} complex)	Red/ orange precipitate	□	□
Chromic Acid Test	Distinguishes aldehydes from ketones	Chromic acid (H_2CrO_4)	Green solution (Cr^{3+} formed from Cr^{6+})	□	□

Carboxylic Acids

Definition and Structure

Carboxylic acids are a class of organic compounds characterized by the presence of the **carboxyl functional group (–COOH)**, which consists of a **carbonyl group (C=O)** directly bonded to a **hydroxyl group (–OH)** on the **same carbon atom**. This combination of functionalities imparts **acidic behavior** and **chemical reactivity** in both biological and laboratory contexts. The **general formula** is **R–COOH**, where *R* represents a hydrogen atom or a hydrocarbon chain.

Carboxylic acids are typically **polar molecules**, capable of **forming hydrogen bonds**, which enhances their **solubility in water**, especially among **low molecular weight acids**. In aqueous solution, they behave as **weak acids**, donating a proton (H^+) to form a **carboxylate ion ($R–COO^-$)**.

These compounds are **widespread in nature** and found in many **biological sources and food products**:

- **Formic acid** – present in ant and bee venom.
- **Acetic acid** – the main acidic component of vinegar.
- **Lactic acid** – produced during **bacterial fermentation** in sour milk and yogurt.
- **Citric acid** – a **tricarboxylic acid** found in citrus fruits (e.g., lemons, oranges), essential in the **Krebs cycle** of cellular respiration.
- **Butyric acid** – produced in the **colon** during fermentation of dietary fiber, contributing to **gut health**.
- **Fatty acids** (e.g., **palmitic acid**, **oleic acid**) – found in **animal fats** and **plant oils**, essential for **cell membrane integrity** and **energy storage**.

The abundance of carboxylic acids in plants, animals, and fermented products reflects their fundamental role in metabolism, nutrition, and health.

IUPAC Naming Rules of Carboxylic Acid

In the IUPAC system, carboxylic acids are named by **replacing the "–e" ending** of the parent alkane with **"–oic acid."** The **carbon atom of the carboxyl group (–COOH)** is always assigned **position number one**, so there is **no need to indicate its position** explicitly in the name.

Examples include:

- $HCOOH$ → Methanoic acid *(common name: formic acid)*
- CH_3COOH → Ethanoic acid *(acetic acid)*
- CH_3CH_2COOH → Propanoic acid *(propionic acid)*
- $CH_3(CH_2)_2COOH$ → Butanoic acid *(butyric acid)*

As the carbon chain increases, the naming continues in a systematic fashion. For **aromatic carboxylic acids**, the name is based on the structure of the aromatic ring bonded to the carboxyl group. For instance:

- **Benzoic acid** has the IUPAC name **benzenecarboxylic acid**.

This systematic approach ensures **clarity and consistency** in naming carboxylic acids, especially when dealing with **substituents or complex molecular structures**, and is essential for both academic and clinical communication.

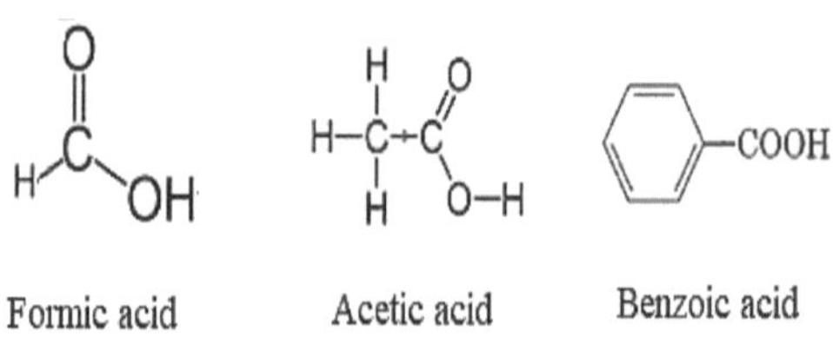

Figure 4.9 Carboxylic Acids–COOH Group

Biological and Medical Significance

Carboxylic acids play a **fundamental role in human physiology and medicine** due to their acidic properties and central involvement in **biochemical and metabolic pathways**. They form the **structural core** of numerous essential biomolecules, including **fatty acids** and **amino acids**, which are critical for **energy production**, **cell membrane integrity**, and **protein biosynthesis**.

In metabolism, simple carboxylic acids such as **lactic acid** and **acetic acid** serve as **intermediates in energy cycles** and help regulate the **acid–base balance** in body fluids. Long-chain fatty acids like **palmitic acid** and **oleic acid** act as **energy stores** and key components of **phospholipid membranes**. **Amino acids**, which contain both an **amino group (–NH_2)** and a **carboxyl group (–COOH)**, rely on the carboxyl group to form **peptide bonds**, the fundamental links in protein structure.

Lactic acid, produced during **anaerobic respiration** in muscles, accumulates under **oxygen-limited conditions**, contributing to **lactic acidosis**, a clinically relevant condition. **Acetic acid**, a simple carboxylic acid, is the active component of **vinegar** and is used for its **antimicrobial properties** in some topical treatments.

Salicylic acid, a naturally occurring **β-hydroxy acid**, is widely used in **dermatology**, particularly for treating **acne**, and serves as the chemical precursor of **aspirin (acetylsalicylic acid)**. In pharmacology, the **carboxyl group's polarity and reactivity** enable molecules to interact with **enzymes, receptors, and transport proteins**, affecting **drug solubility, absorption**, and **bioavailability**.

Several **nonsteroidal anti-inflammatory drugs (NSAIDs)**, including **ibuprofen**, contain carboxylic acid groups that are essential for their **anti-inflammatory and analgesic activity**.

In summary, carboxylic acids are **chemically versatile and biologically indispensable**, with broad applications in **health maintenance, disease treatment**, and **drug development**.

Table 4.10: Biologically and Clinically Significant of Carboxylic Acids

Compound	Biological / Clinical Role
Acetic acid	Metabolic intermediate; antimicrobial; used in IV infusions
Lactic acid	Produced during anaerobic respiration; linked to acidosis
Palmitic acid	Saturated fatty acid; energy source and membrane component
Oleic acid	Monounsaturated fatty acid; supports heart and membrane health
Salicylic acid	Topical acne treatment; aspirin precursor
Benzoic acid	Antifungal; food preservative; topical use
Propionic acid	Gut microbiome product; antimicrobial and metabolic roles
Butyric acid	Short-chain fatty acid; promotes gut and colon health

Amino Compounds

Definition and Structure

Amino compounds are a class of organic molecules defined by the presence of an **amino group (–NH_2)**, in which a **nitrogen atom** is covalently bonded to one or more **hydrogen atoms** and attached to a **carbon skeleton**. This nitrogen-containing group is central to many biologically active molecules and is particularly important in **amino acids**, the fundamental building blocks of **proteins**.

The **amino group** functions as a **weak base** due to the **lone pair of electrons on the nitrogen atom**, enabling it to accept a **proton (H^+)**

and engage in **acid–base reactions**, **hydrogen bonding**, and **nucleophilic substitution** reactions. The simplest form of an amino compound is a **primary amine**, with the general formula **R–NH_2**, where *R* represents an **alkyl** or **aryl** group.

The **structure and position** of the amino group significantly affect the compound's **chemical reactivity**, **solubility**, and **interaction with biological molecules**. Lower molecular weight amines can **form hydrogen bonds** with water, making them **soluble in aqueous environments** such as **blood plasma**. Their **basic nature** allows them to react with acids to form **salts**, a feature frequently used in **drug formulation** to enhance **solubility** and **bioavailability**.

Biologically, the amino group is crucial for:

- **Peptide bond formation** in **protein synthesis**
- The activity of **neurotransmitters** like **dopamine** and **serotonin**
- The function of **synthetic drugs** such as **antihistamines** and **local anesthetics**

In **clinical chemistry and pharmacology**, a deep understanding of amino compounds is essential for interpreting **metabolic pathways**, designing **therapeutic agents**, and predicting **drug behavior** within the human body.

Classification of Amines

Amines are classified based on the number of carbon-containing (organic) groups attached to the nitrogen atom, a distinction that significantly influences their chemical properties, reactivity, and biological roles.

Primary amines (1°) contain one alkyl or aryl group bonded to the nitrogen atom along with two hydrogen atoms. They follow the general formula **R–NH_2** and include simple compounds such as **methylamine (CH_3–NH_2)**. Biologically, primary amines are commonly found in **amino acids** and **small signaling molecules**, playing fundamental roles in metabolism and protein synthesis.

Secondary amines (2°) have two organic groups (alkyl or aryl) attached to the nitrogen atom, along with one hydrogen. Their general structure is **R–NH–R′**, and an example is **dimethylamine (CH_3–NH–CH_3)**. These amines are frequently encountered in **pharmaceutical formulations** and **metabolic intermediates**, where their reactivity contributes to drug activity or transformation.

Tertiary amines (3°) contain three organic groups bonded to the nitrogen atom, with no hydrogen atoms directly attached to nitrogen. They are represented by the general formula **R–N(R′)–R″**, with **trimethylamine ($N(CH_3)_3$)** as a typical example. Tertiary amines appear in important **biological molecules such as neurotransmitters** and are components of certain **local anesthetics** and drug carriers.

Quaternary ammonium compounds (4°) feature a nitrogen atom bonded to four organic groups and bear a **permanent positive charge (R_4N^+)**. A well-known example is **choline**, a precursor to **acetylcholine**, which is essential for **nerve impulse transmission**, **membrane structure**, and **drug transport** across biological barriers.

Understanding these classifications is essential for explaining the **acid–base behavior**, **solubility**, and **interaction** of amines with biological targets such as **enzymes** and **receptors**. In medicinal chemistry, this framework aids in designing drugs that mimic or inhibit the action of **amine-based biomolecules**, enabling the development of effective therapeutic agents.

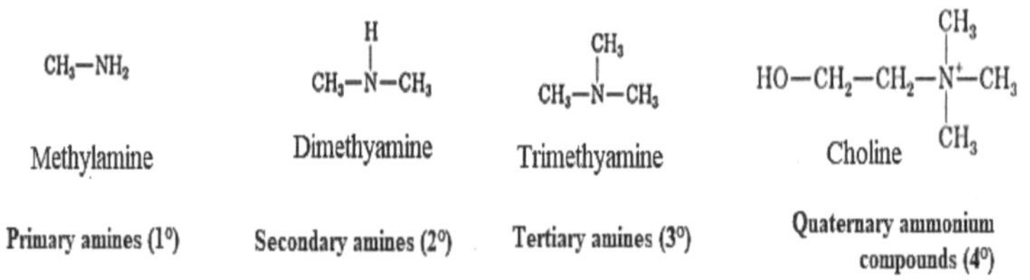

Figure 4.10 Classification of Amines

Biological and Medical Significance of Amino Compounds

Amino compounds are essential to life due to their involvement in a wide range of **biological functions** and **medical applications**. One of the most prominent examples is their presence in **amino acids**, which are the fundamental building blocks of proteins. Proteins are critical for **tissue construction**, **enzymatic catalysis**, and the **regulation of physiological processes** such as immunity and hormonal signaling.

The **amino group (–NH_2)** is also a key component of many **neurotransmitters**, including **dopamine**, **serotonin**, and **histamine**, which regulate **brain function**, **mood**, **sleep**, and **immune responses**. In clinical medicine, numerous drugs and therapeutic agents are either **derived from** or **designed to mimic natural amines**. Examples include **epinephrine (adrenaline)** and **norepinephrine (noradrenaline)**, which are critical for the **stress response**, and **acetylcholine**, a **quaternary amine** essential for **nerve impulse transmission** and **muscle control**.

The **basic nature** of the amino group allows these compounds to interact with **acids**, **receptors**, and **enzymes**, making them **chemically versatile** and **biologically indispensable** in both **diagnostic** and **therapeutic** settings. Many **pharmaceutical agents**, such as **antihistamines**, **antidepressants**, and **local anesthetics**, include amine groups to enable targeted interaction with biological systems.

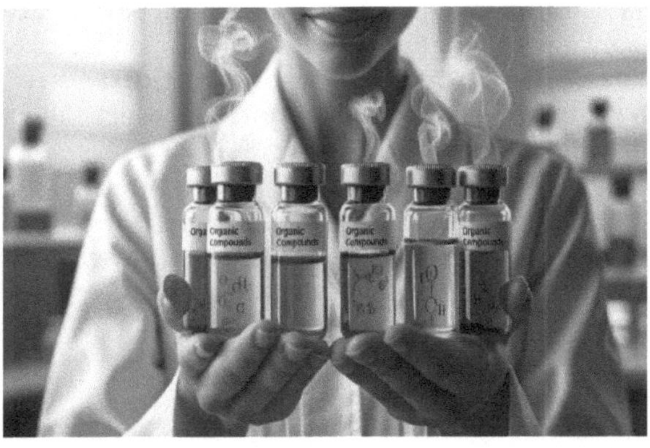

Table 4.11 Biologically and Clinically Significant of Amino Compounds

Compound	Functional Group	Biological Role
Amino acids	$-NH_2 + -COOH$	Protein synthesis, enzyme function
Dopamine	Catecholamine amine	Neurotransmitter: mood, reward system
Histamine	Heterocyclic amine	Immune response, allergic reaction
Epinephrine	Amine hormone	Stress response (fight-or-flight)
Acetylcholine	Quaternary ammonium amine	Nerve impulse transmission, muscle contraction

Other Organic Compounds of Biomedical Interest

Beyond the primary classes of organic compounds—**alcohols, aldehydes, ketones, carboxylic acids**, and **amines**—there exists a diverse range of **other organic molecules** that are critically important to **biological systems** and **medical applications**. These include **esters**, **ethers**, **phenols**, **amides**, **thiols**, **nitriles**, and **organophosphates**, each with distinct structural features and functional roles in **biochemistry** and **clinical therapeutics**.

Esters, formed from the reaction between a **carboxylic acid** and an **alcohol**, are known for their **fragrant aromas** and their central role in **lipid metabolism**, particularly in **triglycerides**. Esters are also vital in **drug design**—for example, **aspirin**, in which the ester group is key to its **anti-inflammatory activity**.

Ethers, characterized by an **oxygen atom** linking two carbon chains (R–O–R′), are used in **pharmaceutical preparations** and historically served as **general anesthetics**, such as **diethyl ether**.

Phenols contain a **hydroxyl group (–OH)** bonded to an **aromatic ring**, giving them enhanced **acidity** and **reactivity** compared to alcohols. Phenols exhibit strong **antiseptic** and **antioxidant** properties, making them useful in **oral care** and **topical antimicrobial agents**.

Amides, formed by the condensation of a **carboxylic acid** and an **amine**, form the **peptide bonds** that link amino acids in **proteins**. These bonds maintain the **secondary and tertiary structure** of proteins, essential for their **function and stability**.

Thiols (–SH), structurally similar to alcohols, exhibit distinct chemical behavior. They are essential in forming **disulfide bridges** (as in **cysteine**), which stabilize the **three-dimensional conformation** of proteins such as **insulin**. Thiols also play a role in **antioxidant defense**, as seen in **glutathione**.

Nitriles (–C≡N), commonly found in **pharmaceutical intermediates**, contribute to **metabolic pathways** and **drug synthesis**. Their linear structure and **chemical reactivity** make them valuable in the design of **bioactive molecules**.

Organophosphates, containing **phosphorus atoms bonded to oxygen and carbon**, are structurally important in molecules like **ATP** and **DNA**. They are vital for **energy transfer** and **genetic information storage**, though some synthetic derivatives can act as **potent neurotoxins** by inhibiting **acetylcholinesterase**, thereby affecting **nerve signaling**.

These compounds illustrate the **chemical diversity** underpinning life processes and therapeutic innovation. Their structural and functional versatility underscores the importance of **organic chemistry** in modern **nursing, clinical practice**, and **drug development**.

Table 4.12 **Other Organic Compounds of Biomedical Relevance**

Compound Class	**Functional Group**	**Key Examples**	**Biological / Medical Relevance**
Esters	–COOR	Aspirin, Triglycerides	Involved in fat metabolism and drug formulation
Ethers	R–O–R′	Diethyl Ether	Historically used as anesthetic and solvent
Phenols	Aromatic ring with –OH	Thymol, Eugenol	Antiseptic and antioxidant in oral and topical care
Amides	$–CONH_2$	Peptide bonds in proteins	Backbone of proteins, essential for enzyme activity
Thiols	–SH	Cysteine, Glutathione	Stabilize proteins and antioxidant defense
Nitriles	–C≡N	Acetonitrile	Used in drug synthesis and metabolic processes
Organophosphates	Phosphorus with ester linkages	DNA backbone, Acetylcholine analogs	Crucial for genetics and neurotransmission; some neurotoxic

Chemistry Link to Health

Bioorthogonal Chemistry:

A New Frontier for In Vivo Diagnostics and Targeted Therapies

Recent advancements in bioorthogonal chemistry—a field focused on chemical reactions that occur inside living organisms without interfering with native biological processes—have revolutionized biomedical research and clinical applications. Central to this innovation are organic functional groups such as azides ($–N_3$), strained alkynes, and aldehyde/hydrazone pairs, which can selectively undergo "click reactions" in vivo. These reactions enable precise labeling of biomolecules, targeted imaging of tumors, and controlled drug release directly at disease sites. For example, pretargeted PET imaging uses azide-tagged antibodies that localize to tumors, followed by injection of a radioactive probe bearing a cyclooctyne group, allowing the two to react and generate a clear, high-resolution image. Similarly, smart therapeutics can be activated only at the tumor site by a click-to-release mechanism, minimizing side effects and enhancing treatment specificity. These cutting-edge applications showcase how fundamental principles of organic chemistry—especially the reactivity and biocompatibility of functional groups—are driving a new era of personalized medicine, noninvasive diagnostics, and safer therapeutic interventions. For health professionals, understanding these chemical strategies opens the door to interpreting emerging technologies in oncology, imaging, and drug delivery.

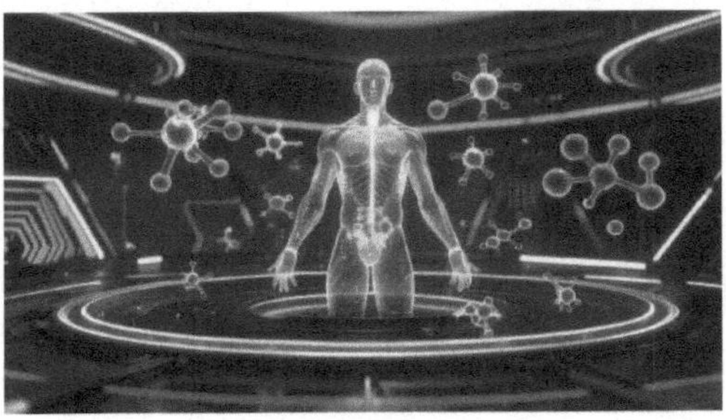

Chapter summary

This chapter explored the foundational classes of organic molecules that are essential to human health, biological function, and medical practice. Emphasis was placed on the structure, nomenclature, chemical behavior, and biomedical relevance of key functional groups and compound classes.

Organic reactivity is primarily governed by the presence of specific functional groups such as hydroxyl (–OH), carbonyl (C=O), carboxyl (–COOH), and amino ($-NH_2$). These groups are responsible for the chemical identity and behavior of biomolecules, enabling them to participate in vital physiological reactions. For instance, the hydroxyl group contributes to solubility in alcohols and sugars, while the carboxyl and amino groups are essential in the formation of proteins and buffering systems in the body.

Hydrocarbons represent the simplest class of organic compounds, consisting solely of carbon and hydrogen. They are classified into alkanes (single bonds), alkenes (double bonds), alkynes (triple bonds), and aromatic compounds (with delocalized electrons in rings). The degree of saturation in hydrocarbons influences their physical properties and biological effects—saturated fats are solid at room temperature and may raise cholesterol, while unsaturated fats are typically liquid and considered heart-healthier.

Alcohols are organic compounds containing one or more hydroxyl (–OH) groups. They serve diverse functions in healthcare, ranging from ethanol used as a disinfectant to glycerol found in lipid metabolism. Alcohols are classified based on the number of hydroxyl groups into monohydric, dihydric, or trihydric forms. Their ability to form hydrogen bonds makes them highly polar and soluble in water, which enhances their utility in both biological systems and pharmaceutical formulations.

Aldehydes and ketones are characterized by the carbonyl functional group (C=O), but differ in its placement. In aldehydes, the carbonyl group is located at the end of the carbon chain (e.g., formaldehyde, retinal), while in ketones, it is found within the chain (e.g., acetone, ketone bodies). These compounds are crucial in metabolism and clinical diagnostics. For example, ketones are elevated in diabetic ketoacidosis, while retinal, an aldehyde, plays a key role in the visual cycle.

Carboxylic acids possess a carboxyl group (–COOH) that imparts acidic properties and participates in hydrogen bonding and metabolic regulation. They are widely found in biological molecules such as fatty acids and amino acids and are also used therapeutically—salicylic acid, for instance, is the precursor of aspirin. Carboxylic acids help regulate physiological pH, influence drug solubility and bioavailability, and are central to energy metabolism pathways.

Amino compounds, or amines, contain one or more amino groups (–NH_2) and are structurally diverse, ranging from primary to quaternary amines. They are fundamental to life, forming the core structure of amino acids, neurotransmitters, and many therapeutic agents. Amines influence protein synthesis, hormone production, and brain function, making them indispensable in both health and medicine. Their basicity and reactivity also enable them to participate in ionic interactions and enzymatic catalysis within the body.

Beyond the primary classes, several other organic compounds are of major biomedical importance. These include esters, ethers, phenols, amides, thiols, nitriles, and organophosphates—each with distinct structural and functional roles in biochemistry and therapeutics. Esters participate in lipid metabolism and drug design, as in aspirin. Ethers have pharmaceutical uses and once served as anesthetics. Phenols exhibit antiseptic and antioxidant properties. Amides form peptide bonds in proteins. Thiols contribute to protein structure via disulfide bridges. Nitriles are used in drug synthesis, and organophosphates are critical in energy transfer (e.g., ATP) and genetics (e.g., DNA), though some synthetic forms act as neurotoxins. These compounds exemplify the chemical diversity essential to life and highlight the critical role of organic chemistry in nursing, clinical care, and pharmaceutical development.

EXERCISES

Multiple Choice

1. Which of the following functional groups is found in all amino acids?
   A. –OH B. –CHO C. –COOH and $–NH_2$ D. –C=O
2. Acetone is classified as a:
   A. Ketone B. Alcohol C. Aldehyde D. Ester
3. Which compound contains a carboxylic acid group?
   A. Glycerol B. Acetic acid C. Acetone D. Methanol
4. The carbonyl group in a ketone is:
   A. At the end of the chain
   B. Attached to a nitrogen atom
   C. Between two carbon atoms
   D. In the hydroxyl group
5. Which functional group is primarily responsible for the acidic nature of fatty acids?
   A. $–NH_2$ B. –OH C. –COOH D. –C=O

Short Answer Questions

1. Define a functional group and explain its importance in organic molecules.
2. Describe the structural difference between an aldehyde and a ketone.
3. Give one example of a monohydric, dihydric, and trihydric alcohol and mention one medical use for each.
4. How do ketone bodies form in the human body, and in what clinical condition are they elevated?
5. Explain the role of carboxylic acids in maintaining blood pH.

Clinical Application

1. A patient with uncontrolled diabetes has a fruity odor on their breath. Explain the chemical basis for this symptom.
2. Why are amino compounds important in the synthesis of neurotransmitters? Provide one example.

3. Describe how knowledge of organic functional groups helps in understanding the action of a common drug such as aspirin.
4. A burn wound is being disinfected with ethanol. What functional group is responsible for its antiseptic property?
5. Explain how the presence of a carboxylic acid group in a drug affects its solubility and absorption in the body.

Complete the Sentence

1. The functional group that defines alcohols is the __________ group.
2. Ketones differ from aldehydes in that the carbonyl group in ketones is located __________.
3. An example of a trihydric alcohol used in pharmaceutical preparations is __________.
4. In the body, carboxylic acids act as weak __________, helping maintain pH balance.
5. Amino acids contain both __________ and __________ functional groups.
6. The simplest aldehyde, used as a tissue preservative, is called __________.
7. The compound responsible for the fruity breath odor in diabetic ketoacidosis is __________.
8. Esters are formed from the reaction between a(n) __________ and a(n) __________.

True or False

1. ___ All carboxylic acids contain both a carbonyl and hydroxyl group on the same carbon.
2. ___ Acetone is an example of a primary alcohol.
3. ___ Formaldehyde is toxic but is used in medical disinfection.
4. ___ Amino acids contain at least one amino group and one carboxyl group.
5. ___ Ketone bodies are produced only in cases of poisoning.
6. ___ Esters are commonly found in natural fats and oils.
7. ___ The $-NH_2$ group is acidic in nature.
8. ___ Glycerol contains three hydroxyl groups and is classified as a trihydric alcohol.

9. ___ The carbonyl group (C=O) is found in both aldehydes and ketones.
10. ___ Phenol is a type of alcohol with antiseptic properties due to its aromatic ring.

Chapter 5

Introduction to Biochemistry for Healthcare

Chapter 5:Introduction to Biochemistry for Healthcare

Learning Objectives

By the end of this chapter, you should be able to:

1. Identify the structure, types, and functions of biomolecules.
2. Explain key metabolic pathways and energy production.
3. Distinguish between essential, non-essential,amino acids .
4. Interpret biochemical markers in lab tests relevant to patient care.
5. Recognize metabolic disorders .
6. Apply biochemical knowledge in nursing practice.

Biochemistry is the scientific discipline that investigates the chemical processes and substances occurring within living organisms. Serving as a bridge between biology and chemistry, it focuses on the structure, function, and interactions of key biomolecules such as proteins, lipids, carbohydrates, and nucleic acids. The field encompasses a wide range of topics, from the fundamental mechanisms of cellular activity to the molecular basis of disease.

For nursing students, biochemistry is indispensable. It enables the interpretation of laboratory values, fosters an understanding of biochemical alterations in various disease states, and informs clinical decision-making. This knowledge enhances nursing practice by elucidating how the body maintains homeostasis, responds to pharmacological interventions, and displays biochemical indicators during illness. Additionally, biochemistry intersects with physiology by explaining the chemical basis of normal body functions, and with pathology by revealing deviations from these processes in disease. Mastery of biochemistry equips nurses to deliver evidence-based care, educate patients effectively, and contribute meaningfully within multidisciplinary healthcare teams.

Carbohydrates

Carbohydrates are essential biomolecules composed of carbon, hydrogen, and oxygen, usually in the ratio of 1:2:1. As one of the four major classes of biomolecules—alongside lipids, proteins, and nucleic acids—carbohydrates play a central role in sustaining life by providing energy, supporting metabolic regulation, and enabling cellular communication. They are categorized based on the number of sugar (saccharide) units:

- **Monosaccharides** (e.g., glucose, fructose) are the simplest sugars and serve as the fundamental building blocks.
- **Disaccharides** (e.g., sucrose, lactose) consist of two monosaccharide units.
- **Polysaccharides** (e.g., glycogen, starch, cellulose) are complex carbohydrates formed by many sugar units.

Biologically, carbohydrates are the primary energy source for cells, especially for the brain and red blood cells. Clinically, monitoring glucose levels—a key monosaccharide—is vital in diagnosing and managing diabetes mellitus. Glycogen, a storage polysaccharide found in the liver and muscles, plays a crucial role in maintaining blood glucose during fasting or physical activity. Disorders in carbohydrate metabolism can lead to significant health issues such as hypoglycemia, hyperglycemia, or glycogen storage diseases. For nurses, a clear understanding of carbohydrate structure and function supports effective nutritional assessment, diabetes management, and the recognition of symptoms related to metabolic imbalances.

What Are Carbohydrates?

Carbohydrates are **organic biomolecules** that consist primarily of **carbon (C)**, **hydrogen (H)**, and **oxygen (O)**, usually in a hydrogen-to-oxygen ratio of **2:1**, similar to water. They are often represented by the empirical formula $(CH_2O)_n$ or $Cm(H_2O)_n$, where *m* and *n* may differ depending on the compound.

Chemically, carbohydrates are characterized by the presence of a **carbonyl group** (either an **aldehyde** or a **ketone**) and multiple **hydroxyl groups (–OH)**. While most carbohydrates conform to the general formula, some exceptions exist—such as **uronic acids**, which deviate from this ratio, and compounds like **formaldehyde** or **acetic acid**, which resemble carbohydrates structurally but are not classified as such.

Classification of Carbohydrates

Carbohydrates are classified based on the number of sugar units they contain and their ability to undergo hydrolysis. The simplest form is **monosaccharides**, which consist of a single sugar unit and cannot be further hydrolyzed. They include glucose, fructose, and galactose, and are categorized further by the number of carbon atoms (trioses, pentoses, hexoses) and the type of carbonyl group they contain (aldoses or ketoses).

Disaccharides are formed by the condensation of two monosaccharide units through a glycosidic bond; common examples include sucrose (glucose + fructose), lactose (glucose + galactose), and maltose (glucose + glucose). These sugars can be hydrolyzed into their monosaccharide components and are important dietary sources of energy.

Moving further in complexity, **oligosaccharides** contain between three and ten monosaccharide units. They are commonly found attached to proteins and lipids in the body, where they play roles in cell recognition, signaling, and immune response.

Lastly, **polysaccharides** are long chains of monosaccharides, typically more than ten units, and may be linear or branched. They are divided into **homopolysaccharides**, like starch, glycogen, and cellulose (composed entirely of glucose), and **heteropolysaccharides**, such as hyaluronic acid and heparin, which contain more than one type of monosaccharide. Polysaccharides serve structural functions (e.g., cellulose in plants) and storage functions (e.g., glycogen in animals), and are vital to energy metabolism and cellular architecture.

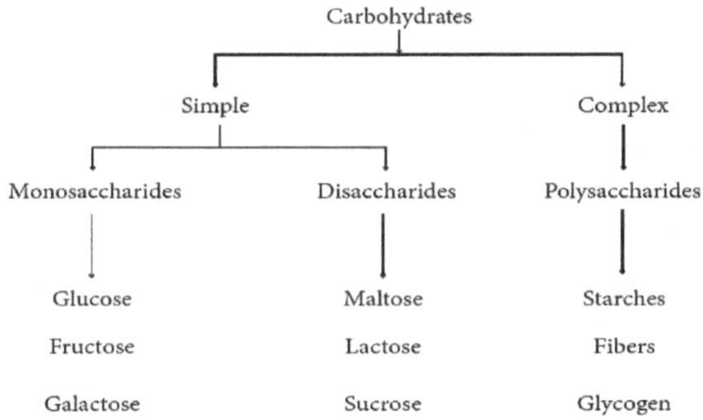

Figure 5.1 Classification of Carbohydrates

Monosaccharaides *(Simple Sugars)*

Monosaccharaides are the simplest form of carbohydrates, consisting of a single sugar unit that cannot be hydrolyzed into smaller units. They serve as the fundamental building blocks for disaccharides and polysaccharides. Structurally, monosaccharaides are typically colorless, crystalline solids that are soluble in water and insoluble in nonpolar solvents. They follow the general chemical formula $\mathbf{C_nH_{2n}O_n}$, with the simplest examples including **dihydroxyacetone** and **glyceraldehyde** (n = 3).

Monosaccharaides are classified based on two main criteria:

- **Type of functional group:**
  - **Aldoses** – contain an aldehyde group (–CHO), e.g., *glucose*
  - **Ketoses** – contain a ketone group (C=O), e.g., *fructose*
- **Number of carbon atoms:**
  - **Trioses** – 3 carbon atoms
  - **Tetroses** – 4 carbon atoms
  - **Pentoses** – 5 carbon atoms
  - **Hexoses** – 6 carbon atoms
  - **Heptoses** – 7 carbon atoms

A prime example is **D-glucose**, the most biologically important monosaccharide. It is classified as an **aldohexose**, containing six carbon atoms and an aldehyde group. D-glucose exists in both open-chain and cyclic forms and plays a central role in **cellular respiration and energy metabolism**. Despite sharing the same molecular formula **$C_6H_{12}O_6$**, monosaccharaides such as glucose, fructose, and galactose differ in the arrangement of their atoms—especially the position of functional groups and asymmetric carbons—resulting in distinct **chemical properties** and **biological functions**.

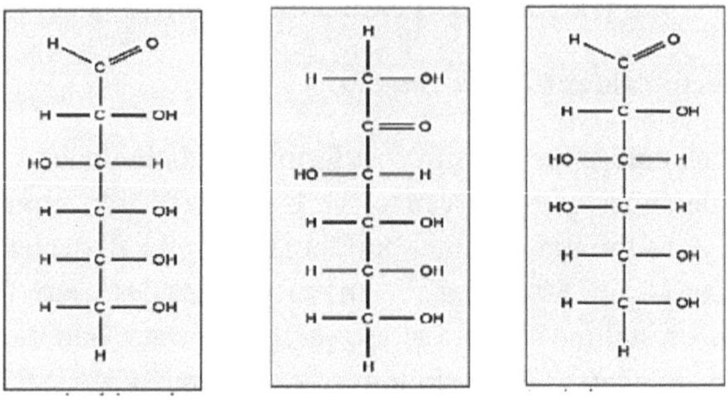

Glucose Fructose Galactose

Figure 5.2 Examples of Monosaccharaides

Functions of Monosaccharaides

Monosaccharides play critical roles in both **metabolism** and **cell structure**, contributing to a wide range of biological functions in living organisms. These simple sugars are not only primary energy sources but also fundamental building blocks for more complex biomolecules.

Glucose ($C_6H_{12}O_6$) is the most important and widely utilized monosaccharide. It serves as the **primary energy source** for both plants and animals. In plants, glucose is synthesized via **photosynthesis** and stored as **starch**. When consumed by animals and humans, it undergoes cellular respiration to produce ATP.

Fructose, a six-carbon **ketose sugar**, is naturally found in **fruits and honey**, where it contributes to sweetness and acts as a **rapidly metabolizable energy source**.

Galactose is particularly important in **infant nutrition**, as it is a component of **lactose**—the main sugar in milk. It supports **brain development** and early metabolic needs.

Ribose, a five-carbon **pentose sugar**, forms the structural backbone of **RNA** and several essential coenzymes, such as **ATP**, **NAD^+**, and **FAD**, which are involved in **energy transfer** and **oxidation-reduction reactions**.

Mannose plays a key role in the biosynthesis of **glycoproteins** and **mucoproteins**, essential for **cell signaling**, **immune response**, and **intercellular communication**.

Together, these monosaccharides support **critical physiological functions** across all life forms, from energy metabolism to genetic expression and cellular interactions.

Disaccharides

Disaccharides are carbohydrates composed of **two monosaccharide units** joined by a **glycosidic bond**, a type of **covalent linkage** formed via a **condensation reaction**, which releases one molecule of water. These sugars are typically **sweet-tasting**, **water-soluble**, and play crucial roles in **nutrition**, **energy metabolism**, and **digestive physiology**.

Common Dietary Disaccharides:

Sucrose: Composed of glucose and fructose, it is commonly known as **table sugar**. It is the **primary transport form of carbohydrate in plants** and a major source of dietary sugar in humans.

Lactose: Made of glucose and galactose, **lactose** is the **main sugar in milk**. It is critical for **infant nutrition**. A deficiency in the enzyme **lactase** leads to **lactose intolerance**, resulting in digestive symptoms such as bloating, gas, and diarrhea.

Maltose: Consists of two glucose units. It is formed during the **enzymatic hydrolysis of starch**, especially during digestion and in germinating grains. Maltose serves as an **intermediate in carbohydrate metabolism**.

Trehalose: Composed of two glucose molecules linked by an **α(1→1)** bond. It is found in **mushrooms, insects, and yeast**, and is hydrolyzed by the enzyme **trehalase**.

Cellobiose: Formed from two glucose units linked by a **β(1→4)** bond. It results from the **partial hydrolysis of cellulose** and is **not digestible by humans**, but fermentable by certain microbes.

Each disaccharide must be **hydrolyzed by a specific digestive enzyme** before the **monosaccharide components** can be absorbed into the bloodstream. Understanding their structure and digestion is important in **clinical nutrition, metabolic disease diagnosis**, and the management of **dietary carbohydrate disorders**.

Table 5.1 Comparison of Common Disaccharides

Disaccharide	**Monosaccharide Components**	**Glycosidic Linkage**	**Digestive Enzyme**	**Dietary Source**
Sucrose	Glucose + Fructose	α(1→2)	Sucrase	Table sugar, fruits, vegetables
Lactose	Glucose + Galactose	β(1→4)	Lactase	Milk and dairy products
Maltose	Glucose + Glucose	α(1→4)	Maltase	Germinating grains,
Trehalose	Glucose + Glucose	α(1→1)	Trehalase	Mushrooms, insects
Cellobiose	Glucose + Glucose	β(1→4)	Cellobiase (in microbes)	Product of cellulose breakdown (not digestible by humans)

Reducing vs. Non-Reducing Disaccharides

Disaccharides are classified as **reducing** or **non-reducing sugars** based on the **presence or absence of a free hemiacetal or hemiketal group** on one of their monosaccharide units.

Reducing Disaccharides

Reducing disaccharides contain a **free anomeric carbon** on at least one monosaccharide unit that is **not involved** in the glycosidic bond. This free carbon allows the molecule to adopt a **linear form**, enabling it to **donate electrons** and act as a **reducing agent**. Examples include: **Maltose** and **Lactose**.

These sugars can be detected using **Benedict's or Fehling's solution**, which produce a color change upon reduction, making them useful in **clinical tests for glucose and other reducing sugars**.

Non-Reducing Disaccharides

In contrast, non-reducing disaccharides, such as **sucrose**, have **both anomeric carbons involved** in the glycosidic linkage. This structure **prevents ring opening** and thus **eliminates reducing capability**. As a result: **Sucrose does not react** with Benedict's or Fehling's solution. This distinction is important in **biochemistry, food science**, and **clinical diagnostics**.

Understanding whether a disaccharide is reducing or non-reducing has implications for **laboratory testing**, **digestion**, and the **chemical behavior** of sugars in biological systems.

Functions of Disaccharides

Disaccharides serve a variety of essential biological and nutritional roles in both plants and animals, depending on their structure and source. **Sucrose**, composed of glucose and fructose, is the primary transport sugar in plants and a major source of dietary energy for humans, commonly consumed as table sugar. **Lactose**, a disaccharide of glucose and galactose, is the principal carbohydrate found in mammalian milk and plays a critical role in infant nutrition by supplying energy and supporting brain development. **Maltose**, consisting of two glucose units, is an intermediate product formed during the enzymatic breakdown of starch and is important in digestive processes. **Trehalose**, also a glucose dimer, functions as a stable energy source in insects and provides protection against dehydration and environmental stress. **Cellobiose** results from the partial hydrolysis of cellulose and is a structural disaccharide not digestible by humans, but important in microbial cellulose degradation. **Gentiobiose**, another glucose-based disaccharide, is found in plant glycosides and contributes to various plant metabolic and defense mechanisms. Understanding these functional roles of disaccharides enhances comprehension of carbohydrate metabolism, nutrition, and ecological interactions across species.

Oligosaccharides (3–10 Sugar Units)

Oligosaccharides are carbohydrates composed of **three to ten monosaccharide units** linked together by **glycosidic bonds**. Unlike simpler sugars, oligosaccharides are not typically found freely circulating in biological systems. Instead, they are most commonly bound to lipids or proteins, forming **glycolipids** and **glycoproteins**, respectively. These conjugated structures play essential roles in **cell recognition**, **immune response**, **hormone signaling**, and **protein stability**. Oligosaccharides contribute to the structural diversity of cell surfaces, enabling cells to communicate with their environment and with each other. Their specific arrangement and branching patterns determine biological specificity and are often exploited by pathogens during infection or by the immune system in targeting abnormal cells.

Types of Oligosaccharide Linkages

Oligosaccharides that are bound to proteins form **glycoproteins**, and the nature of this linkage determines their classification and site of synthesis.

In **N-linked glycosylation**, the oligosaccharide is covalently attached to the **amide nitrogen** of the amino acid **asparagine**. This process begins in the **endoplasmic reticulum**, where a core oligosaccharide is assembled and transferred en bloc to the growing polypeptide chain.

In contrast, **O-linked glycosylation** involves the attachment of oligosaccharides to the **hydroxyl group** of **serine** or **threonine** residues. This type of linkage is synthesized later in the **Golgi apparatus** and often involves the sequential addition of monosaccharide units. Both N-linked and O-linked oligosaccharides contribute to protein folding, stability, and cellular communication, and defects in these processes can result in a variety of congenital and metabolic disorders.

Functions of Oligosaccharides

Oligosaccharides perform several essential functions in biological systems, particularly through their role in **glycoproteins** and **glycolipids** found on cell surfaces. One of their most critical roles is in **cell communication and recognition**, where specific oligosaccharide patterns enable cells to identify one another and interact appropriately—an essential feature in processes like embryonic development, hormone signaling, and tissue organization. In the **immune system**, oligosaccharides contribute to the specificity of **antibodies** and **antigens**, helping the body distinguish self from non-self and mount appropriate immune responses. They also facilitate **cell adhesion** through interactions with carbohydrate-binding proteins known as **lectins**, which are involved in inflammation, infection control, and wound healing. Additionally, oligosaccharides derived from plant sources function as **dietary fiber**, providing health benefits such as improved digestive health, modulation of gut microbiota, and reduction of cholesterol levels. These diverse roles underscore the importance of oligosaccharides in both cellular function and overall health.

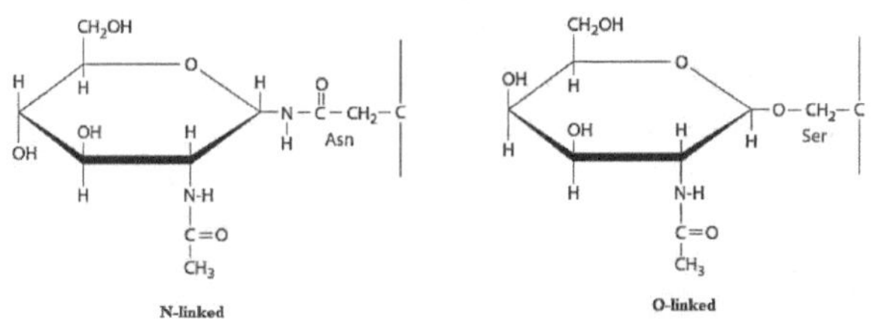

Figure 5.3 Types of Oligosaccharide Linkages

Polysaccharides (More than 10 Monosaccharide Units)

Polysaccharides are large, complex carbohydrates composed of more than ten monosaccharide units joined together by glycosidic bonds. They can range from a few hundred to several thousand sugar residues and exhibit diverse structures and functions depending on the type of monosaccharide, the nature of the glycosidic linkage, and the degree of branching.

Polysaccharides are broadly categorized into two types:

Homopolysaccharides: (Composed of One Type of Monomer)

Homopolysaccharides are complex carbohydrates formed by the polymerization of a single type of monosaccharide. Despite their structural simplicity in terms of monomer composition, they perform diverse and essential roles in both plants and animals.

• **Cellulose**
Cellulose is composed of repeating units of **β-D-glucose** linked by **β(1→4) glycosidic bonds**. It forms straight, rigid chains that bundle into fibers, providing structural support to plant cell walls. Humans lack the enzyme **cellulase** required to break these β-linkages, making cellulose indigestible. However, it remains important in human nutrition as **dietary fiber**, promoting gastrointestinal motility and preventing constipation.

• **Starch**
Starch is the principal **storage polysaccharide in plants**, composed of **α-D-glucose** units. It exists in two molecular forms:

- **Amylose**: A linear polymer with **α(1→4)** linkages. It adopts a helical structure and is slowly digested.
- **Amylopectin**: A highly branched polymer with **α(1→4)** linkages in the linear chains and **α(1→6)** linkages at the branch points. Its branched structure allows for rapid enzyme access and energy release.

Starch is hydrolyzed in the human digestive system by enzymes such as **amylase**, releasing glucose for energy.

• Glycogen
Glycogen is the **animal storage form of glucose**, structurally similar to amylopectin but even more extensively branched. It consists of **α(1→4)** glycosidic bonds in the chains and **α(1→6)** bonds at branch points every 8–12 glucose units. Glycogen is stored primarily in the **liver and skeletal muscles**, serving as a readily mobilized energy reserve. Liver glycogen helps regulate blood glucose levels, while muscle glycogen provides fuel during physical activity.

• Chitin
Chitin is a structural polysaccharide found in the **exoskeletons of arthropods** (e.g., insects, crustaceans) and the **cell walls of fungi**. It is composed of **N-acetyl-D-glucosamine** units linked by **β(1→4)** bonds, resembling cellulose in structure but containing an acetylated amino group. Chitin is not digestible by humans, but it plays a significant ecological role and is used in medical applications such as wound dressings and biodegradable sutures.

Heteropolysaccharides: (More Than One Type of Monomer)

Heteropolysaccharides are complex carbohydrates composed of **two or more different types of monosaccharide units**. These polysaccharides are typically found in structural and functional roles, often linked to proteins or other biomolecules to form **glycoproteins**, **proteoglycans**, or **glycolipids**.

They play essential roles in connective tissue structure, lubrication of joints, cell signaling, and immune response.

• **Glycosaminoglycans (GAGs):** are long, unbranched heteropolysaccharide chains made up of repeating disaccharide units. Each unit typically consists of an **amino sugar**—such as *N-acetylglucosamine* or *N-acetylgalactosamine*—paired with a **uronic acid**, such as *glucuronic acid* or *iduronic acid.* These components give GAGs a **highly negative charge**, enabling them to bind large amounts of water. As a result, GAGs form **hydrated gels** that provide essential **cushioning, viscosity, and elasticity** to connective tissues. Found throughout the **extracellular matrix**, especially in cartilage, skin, and synovial fluid, GAGs help resist compressive forces, facilitate nutrient diffusion, and support tissue hydration. Their biochemical properties make them crucial in joint lubrication, wound healing, and cellular signaling, as well as in clinical applications such as **anticoagulation (heparin)** and **orthopedic treatments (hyaluronic acid injections)**.

Table 5.2 Examples of GAGs and Their Functions:

GAG	Acidic Sugar	Amino Sugar	Biological Role
Hyaluronic acid	D-glucuronic acid	N-acetylglucosamine	Lubricant in joints, eyes; shock absorber
Chondroitin sulfate	D-glucuronic acid	N-acetylgalactosamine	Provides resistance to compression in cartilage
Heparin	L-iduronic acid	N-sulfoglucosamine	Natural anticoagulant; prevents blood clot formation
Dermatan sulfate	L-iduronic acid	N-acetylgalactosamine	Involved in wound healing and vascular health
Keratan sulfate	D-galactose	N-acetylglucosamine	Found in cornea, cartilage, and intervertebral discs

• **Peptidoglycan**: is a rigid, mesh-like heteropolysaccharide that forms a critical structural component of **bacterial cell walls**. It is composed of alternating units of **N-acetylglucosamine (NAG)** and **N-acetylmuramic acid (NAM)**, which are linked together in long chains. These chains are further cross-linked by short **peptide bridges**, creating a strong, lattice-like structure. This cross-linking gives the bacterial cell wall its **mechanical strength**, allowing it to maintain shape, resist osmotic pressure, and protect the cell from environmental stress. Peptidoglycan is especially important in **Gram-positive bacteria**, where it forms a thick outer layer, and it is the **target of several antibiotics** (such as penicillin), which disrupt cell wall synthesis and lead to bacterial lysis.

- **Agarose and Alginate:** are polysaccharides derived from **marine algae** and are widely used in **biomedical and laboratory applications**. **Agarose** is a neutral polysaccharide extracted from red algae and is primarily used to form **stable, non-toxic gels** in techniques such as **gel electrophoresis**, where it serves as a medium for separating DNA and RNA molecules based on size. Its ability to form clear, reproducible gels makes it essential in molecular biology and diagnostic laboratories. **Alginate**, extracted from brown seaweed, is a negatively charged polysaccharide composed of mannuronic and guluronic acid residues. Due to its **biocompatibility and gel-forming properties**, alginate is widely used in **wound dressings** to maintain a moist healing environment, in **controlled drug delivery systems**, and as a **food thickener** in the pharmaceutical and food industries. Both agarose and alginate exemplify how naturally occurring polysaccharides can be adapted for critical roles in modern science and medicine.

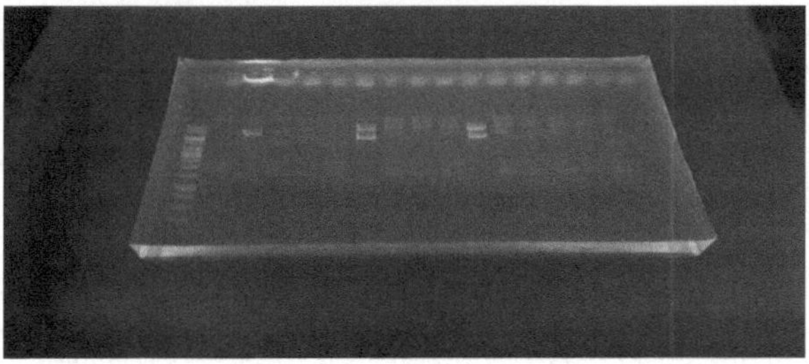

Figure 5.4 Agarose Gel Electrophoresis

Functions of Polysaccharides:

Polysaccharides are large carbohydrate molecules composed of more than ten monosaccharide units, and they perform essential structural, storage, and physiological roles in living organisms. Structurally, polysaccharides such as cellulose, chitin, and peptidoglycan provide rigidity and protection. Cellulose is a major component of plant cell walls, offering strength and acting as dietary fiber in the human digestive system. Chitin, found in the exoskeletons of insects and the cell walls of fungi, serves a similar structural function in non-plant organisms. Peptidoglycan is a complex polysaccharide that reinforces bacterial cell walls and maintains cell shape. In terms of storage, starch serves as the main energy reserve in plants and consists of amylose and amylopectin chains. Glycogen, the animal equivalent, is highly branched and stored in the liver and muscles, allowing for rapid glucose release when energy is needed. Inulin, a plant-based storage polysaccharide, is also important as a source of dietary fiber. Physiologically, certain polysaccharides have specialized biomedical functions. Hyaluronic acid is involved in joint lubrication and tissue hydration. Heparin acts as a natural anticoagulant, preventing unwanted blood clotting in blood vessels. Additionally, chondroitin sulfate and keratan sulfate, found in cartilage, contribute to shock absorption and maintain joint flexibility. Together, these polysaccharides support critical functions in structural biology, metabolism, and medical processes.

Biological Roles and Clinical Relevance of Carbohydrates

Carbohydrates play a fundamental role in the biology of all living organisms and are vital to human health and clinical care. Primarily, they serve as the body's main source of energy, with glucose being the most critical fuel for the brain, muscles, and red blood cells. During periods of increased energy demand or fasting, stored polysaccharides such as glycogen in animals and starch in plants provide a readily accessible energy reserve. Structurally, carbohydrates are essential components of cell membranes; glycoproteins and glycolipids embedded in these membranes participate in cell recognition, communication, and immune defense.

For instance, ABO blood group antigens are defined by specific carbohydrate sequences on red blood cells. Structural polysaccharides like cellulose in plants and chitin in fungi and arthropods contribute mechanical strength and protection. Clinically, disruptions in carbohydrate metabolism are linked to several health conditions. Diabetes mellitus, characterized by impaired glucose regulation, requires careful monitoring of blood sugar and HbA1c levels. Lactose intolerance, caused by a deficiency in the enzyme lactase, results in digestive issues upon consuming milk products. Inherited glycogen storage diseases affect the synthesis or breakdown of glycogen, leading to energy imbalance and tissue damage.

Dietary fiber, composed largely of indigestible polysaccharides like cellulose, promotes gastrointestinal health, lowers cholesterol, and helps regulate blood glucose levels. Carbohydrates also have immunological significance: glycosylation patterns on glycoproteins enable the immune system to distinguish between self and foreign cells, aiding in pathogen defense.

Medically, carbohydrate-based compounds such as heparin (an anticoagulant) and hyaluronic acid (used in joint therapy and cosmetic treatments) highlight their therapeutic relevance. For nurses and allied health professionals, a strong understanding of carbohydrate structure and function enhances clinical decision-making, supports effective patient education, and is essential for managing metabolic, digestive, and inflammatory disorders.

Table 5.3: Biological Roles and Clinical Relevance of Carbohydrates

Category	Function / Role	Clinical Relevance / Examples
Energy Source	Immediate fuel for cells	Glucose – ATP production
Energy Storage	Stores excess glucose	Glycogen, Starch
Structural Function	Structural support for cells	Cellulose, Chitin, Peptidoglycan
Cell Communication	Cell recognition and signaling	Glycoproteins, Glycolipids
Immune Function	Immune response activation	Immune cell glycosylation patterns
Clinical Disorders	Disorders from carbohydrate metabolism	Diabetes, Lactose intolerance, Glycogen storage diseases
Digestive Health	Supports digestion and gut health	Dietary fiber
Pharmaceutical Use	Medicinal carbohydrate uses	Heparin, Hyaluronic acid
Diagnostic Importance	Disease markers and metabolic analysis	Blood glucose, HbA1c, Reducing sugars in urine

Proteins

Proteins are essential macromolecules found in every cell of the human body. They are involved in virtually all biological processes, including **catalysis of metabolic reactions**, **structural support**, **transport of molecules**, **cell signaling**, **immune defense**, and **muscle contraction**. Unlike carbohydrates and lipids, proteins are made from **amino acids**, which are linked in specific sequences to form **polypeptide chains**. These chains fold into unique three-dimensional shapes that determine their biological activity.

Proteins differ greatly in size, shape, and function, depending on the **order of amino acids** and how the chain folds and interacts with other molecules. Some proteins, such as **enzymes**, act as catalysts that accelerate chemical reactions, while others, such as **antibodies**, defend the body against pathogens.

For nursing and health professionals, understanding protein structure and function is vital for interpreting lab results (e.g., serum albumin levels), managing nutritional needs, and recognizing conditions like **protein-energy malnutrition**, **enzyme deficiencies**, and **autoimmune diseases**. Proteins are not only structural and functional elements of life, but also valuable clinical indicators of health and disease.

Amino Acids: Classification, Structure, and Clinical Relevance

There are **20 standard amino acids** found in human proteins, each containing a central carbon atom (called the α-carbon) bonded to an amino group ($-NH_2$), a carboxyl group ($-COOH$), a hydrogen atom, and a unique **side chain (R group)**. The R group determines the chemical nature of each amino acid—whether it is **nonpolar, polar, acidic, or basic**—and directly influences how the amino acid interacts with others, affecting **protein folding, structure, and function**. In physiological conditions, amino acids exist as **zwitterions**—molecules with both positive and negative charges—due to protonation of the amino group and deprotonation of the carboxyl group.

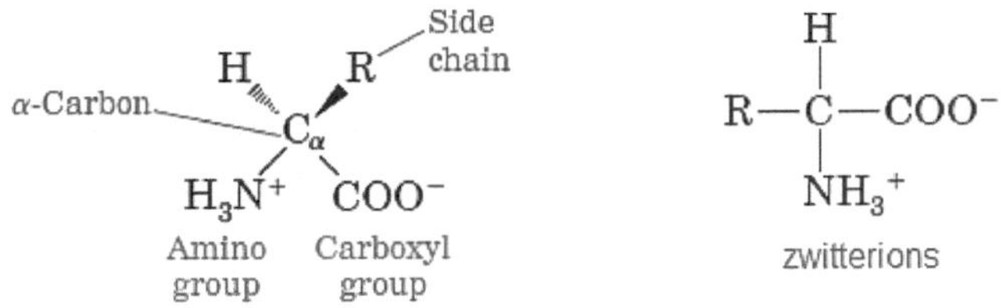

Figure 5.5 General structure of amino acid

Amino acids are broadly classified based on the body's ability to synthesize them. This classification is essential in clinical nutrition, especially for patients who are critically ill, undergoing recovery, or experiencing metabolic stress.

1. Essential Amino Acids

Essential amino acids **cannot be synthesized by the human body** in sufficient amounts and must be obtained from dietary sources. They are vital for growth, tissue repair, immune function, and metabolic regulation.

Table 5.4: Essential Amino Acids and Their Clinical Significance

Amino Acid	Abbreviation	Key Functions / Clinical Notes
Histidine	His	Required for growth, red blood cell production, and histamine synthesis
Isoleucine	Ile	Involved in muscle metabolism, immune response, and hemoglobin formation
Leucine	Leu	Stimulates muscle protein synthesis; helps with tissue regeneration
Lysine	Lys	Supports collagen formation and calcium absorption; essential in wound healing
Methionine	Met	Initiates protein synthesis; precursor to cysteine and antioxidant pathways
Phenylalanine	Phe	Converted to tyrosine; precursor to neurotransmitters like dopamine and epinephrine
Threonine	Thr	Maintains gut mucosa; supports immune function and protein balance
Tryptophan	Trp	Precursor to serotonin and melatonin; influences mood, sleep, and appetite
Valine	Val	Provides energy for muscle tissue; important in recovery and repair

2. Non-Essential Amino Acids

Non-essential amino acids **can be synthesized by the body** and are usually produced in adequate amounts under normal physiological conditions. However, they remain essential for basic metabolic and structural functions.

Table 5.5: Non-Essential Amino Acids and Their Clinical Significance

Amino Acid	Abbreviation	Key Functions / Clinical Notes
Alanine	Ala	Participates in glucose-alanine cycle; important for energy production in muscles
Asparagine	Asn	Important for protein synthesis and nervous system function
Aspartic acid	Asp	Plays a role in the urea cycle and neurotransmitter regulation
Glutamic acid	Glu	Major excitatory neurotransmitter in the central nervous system
Serine	Ser	Involved in metabolism, enzyme activity, and production of phospholipids

3. Conditionally Essential Amino Acids

Conditionally essential amino acids are typically synthesized by the body, but during periods of stress, illness, trauma, or rapid growth, the body's demand may exceed its capacity to produce them. Supplementation may be required in critical care, wound healing, or catabolic states. For example, amino acids such as glutamine and arginine become essential during severe illness or injury because they support immune function and tissue repair. In infants and patients with metabolic disorders, the synthesis of these amino acids may be insufficient, necessitating dietary or intravenous supplementation. Understanding these dynamic needs is crucial in clinical nutrition and nursing care to optimize patient recovery and maintain metabolic balance.

Table 5.6: Conditionally Essential Amino Acids and Their Clinical Significance

Amino Acid	Abbreviation	Key Functions / Clinical Notes
Arginine	Arg	Enhances wound healing, immune response, and nitric oxide production
Cysteine	Cys	Acts as an antioxidant; precursor to glutathione; supports detoxification
Glutamine	Gln	Main fuel for immune cells and intestinal lining; critical during catabolism
Glycine	Gly	Aids in collagen synthesis; inhibitory neurotransmitter in the CNS
Proline	Pro	Essential for collagen stability and tissue regeneration
Tyrosine	Tyr	Precursor for dopamine, norepinephrine, epinephrine, and thyroid hormones

Clinical Insight for Nursing

- **Essential amino acids** must be monitored in malnourished or post-surgical patients, especially in total parenteral nutrition (TPN).
- **Conditionally essential amino acids** like glutamine and arginine are often supplemented in patients with burns, sepsis, or cancer to support healing and immune function.
- Amino acid imbalances can signal metabolic disorders (e.g., **phenylketonuria** due to excess phenylalanine).
- A thorough understanding of amino acid roles allows nurses to interpret lab values, advise on dietary needs, and support patient recovery in diverse clinical scenarios.

Amino acids link together through **peptide bonds**, forming chains called **polypeptides**, which fold into functional proteins. The **sequence** of amino acids (called the **primary structure**) is genetically encoded and crucial to the protein's final structure and biological function.

Peptides

Peptides are short chains of amino acids linked by peptide bonds, a type of covalent bond formed through a condensation (dehydration) reaction. This occurs when the carboxyl group (–COOH) of one amino acid reacts with the amino group ($-NH_2$) of another, releasing a molecule of water (H_2O). The resulting peptide bond has partial double-bond character, which restricts rotation and adds rigidity and structural stability to the peptide chain.

Peptides are classified by the number of amino acid residues:

- Dipeptide: 2 amino acids
- Tripeptide: 3 amino acids
- Oligopeptide: 4 to 10 amino acids
- Polypeptide: more than 10 amino acids

When one or more polypeptide chains fold into a three-dimensional structure, they form a functional protein. This folding is essential for biological activity and enables proteins to serve as enzymes, hormones, antibodies, transporters, and structural components.

For nursing and health professionals, understanding peptide formation is fundamental to topics such as protein synthesis, enzyme function, and clinical biochemistry, including drug action and disease diagnosis.

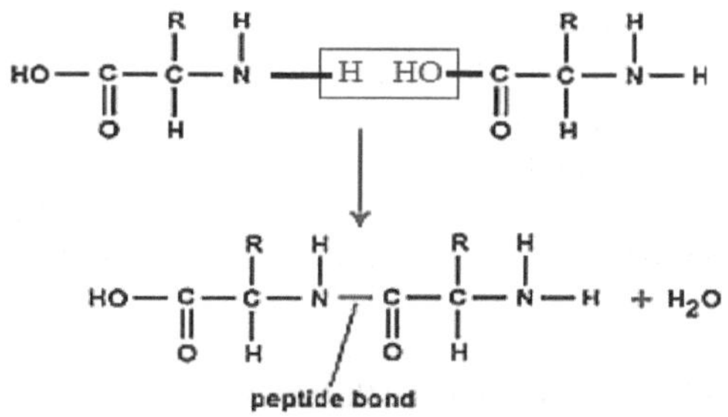

Figure 5.6 formation of peptide bond

Table 5.7: Classification of Peptides

Type of Peptide	Number of Amino Acids	General Description	Example
Dipeptide	2	Formed from two amino acids joined by one peptide bond	Carnosine (found in muscles)
Tripeptide	3	Three amino acids, two peptide bonds	Glutathione (antioxidant)
Oligopeptide	4–10	Short chains with limited function; may act as signaling molecules	Bradykinin (vasodilator peptide)
Polypeptide	>10	Long chain that can fold into parts of a protein or function on its own	Insulin (hormonal polypeptide)
Protein	One or more polypeptides	Folded, functional macromolecule with specific biological roles	Hemoglobin, enzymes, antibodies

Biological and Clinical Importance of Peptides

Peptides play a wide variety of essential roles in human physiology and medicine. Many peptides function as **hormones**, such as **insulin**, which regulates blood glucose levels, and **glucagon-like peptide-1 (GLP-1)** analogs, which are used therapeutically in the management of type 2 diabetes. Others act as **neurotransmitters or neuromodulators**, such as **substance P**, which is involved in pain perception.

Certain peptides also serve as **antibiotics**, such as **bacitracin**, which is used to treat bacterial infections. In clinical diagnostics, **C-peptide** is a key biomarker used to assess endogenous insulin production, helping distinguish between type 1 and type 2 diabetes.

Therapeutic peptides are increasingly used in modern medicine due to their specificity and reduced side effects, making them valuable in areas such as oncology, endocrinology, and immunology. For nurses and healthcare professionals, understanding peptide structure and function is vital for interpreting laboratory results, managing conditions related to **hormonal imbalances, malabsorption, or protein synthesis disorders**, and supporting evidence-based clinical decisions.

Protein Structure and Function

The function of a protein is intimately tied to its **three-dimensional structure**, which is determined by the **specific sequence of amino acids** in its polypeptide chain. Protein structure is organized into **four hierarchical levels** that build upon one another.

The **primary structure** is the unique linear sequence of amino acids joined by **peptide bonds**, directly encoded by the genetic material. This sequence dictates how the protein will fold, and even a **single substitution** (such as in **sickle cell anemia**) can drastically alter protein function and cause disease.

The **secondary structure** refers to localized folding within the chain, stabilized mainly by **hydrogen bonds**. Common patterns include the **alpha (α) helix**, a coiled configuration stabilized by intrachain hydrogen bonding, and the **beta (β) pleated sheet**, formed by hydrogen bonds between adjacent strands. These elements give the protein its **initial shape and mechanical stability**.

The **tertiary structure** describes the overall three-dimensional conformation of a single polypeptide, shaped by interactions among the **R-groups (side chains)** of amino acids. These include **hydrogen bonding**, **ionic interactions**, **disulfide bridges** (covalent S–S bonds), and **hydrophobic interactions**, all of which collectively determine the **active site of enzymes** and the **functional domains** of many proteins.

Some proteins possess a **quaternary structure**, which arises when two or more polypeptide subunits assemble into a larger, functional complex. A classic example is **hemoglobin**, which consists of four subunits—two α and two β chains—that cooperatively bind and transport oxygen in the blood. Understanding these structural levels is essential for nursing professionals to grasp how **protein misfolding**, **enzyme function**, and **drug–receptor interactions** can directly impact patient care.

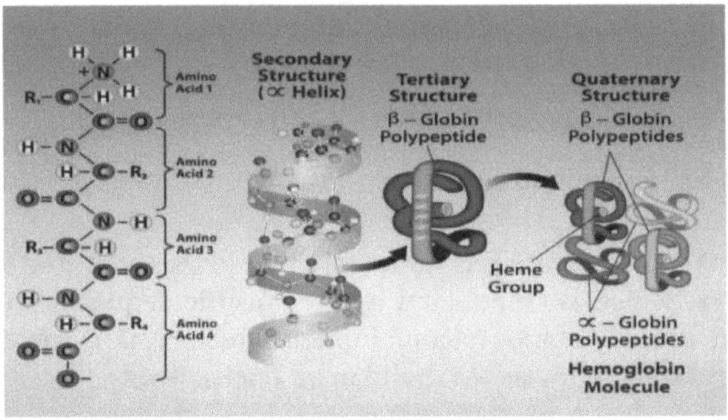

Figure 5.7 Protein Structure

Functions of Proteins and Their Clinical Relevance

Proteins play a central role in nearly every biological function in the human body. Their function is determined by their specific three-dimensional structure and amino acid sequence, which allows them to participate in diverse physiological processes. Some proteins function as **enzymes**, accelerating biochemical reactions essential for digestion and metabolism—for example, *amylase* breaks down starch and *lactase* digests lactose. Others serve **transport roles**, such as *hemoglobin*, which carries oxygen in the blood, and *albumin*, which transports hormones and drugs. **Structural proteins**, such as *collagen* and *keratin*, provide strength and support to tissues like skin, bones, and hair. Certain proteins act as **hormones**, such as *insulin*, which regulates blood glucose levels, while **antibodies** play a critical **immune defense role** by identifying and neutralizing pathogens. Proteins are also involved in **storage functions**, like *ferritin*, which stores iron, and in **movement**, with *actin* and *myosin* facilitating muscle contraction. **Cell signaling proteins**, such as membrane receptors, enable cells to communicate and respond to hormones and environmental stimuli.

Clinically, understanding protein function is vital for nursing practice. Abnormal protein levels can serve as key diagnostic markers. For instance, low *albumin* may indicate malnutrition or liver dysfunction, while elevated liver enzymes like *AST* or *ALT* may suggest hepatic damage. Immunoglobulin patterns help diagnose immune disorders, and genetic mutations affecting protein structure can lead to diseases such as sickle cell anemia or cystic fibrosis. Protein denaturation from fever, acidosis, or toxic exposure may impair function, leading to metabolic disruptions. Therefore, a solid grasp of protein structure and function allows nurses to interpret lab results, educate patients, and support comprehensive and evidence-based care.

Table 5.8 : Biological Functions of Proteins

Function	Example	Biological Role
Enzymatic	*Amylase, Lipase*	Catalyze biochemical reactions (e.g., digestion, metabolism)
Structural	*Collagen, Keratin*	Provide support and strength to tissues such as skin, bones, and hair
Transport	*Hemoglobin, Albumin*	Transport oxygen, hormones, and drugs in the bloodstream
Hormonal	*Insulin, Glucagon*	Regulate physiological processes like glucose metabolism
Immune Defense	*Antibodies (Immunoglobulins)*	Identify and neutralize pathogens
Storage	*Ferritin, Casein*	Store vital substances such as iron.
Movement	*Actin, Myosin*	Involved in muscle contraction and intracellular transport
Cell Signaling	*Receptors, Growth Factors*	Mediate cell responses to hormones.

Protein Denaturation and Clinical Relevance

Protein denaturation refers to the loss of a protein's functional three-dimensional structure due to the disruption of its secondary, tertiary, or quaternary conformation. This process can be triggered by factors such as **high temperature**, **extreme pH changes**, **radiation**,

alcohol, **heavy metals**, or **toxic chemicals**. Although the **primary structure**—the amino acid sequence—remains unchanged, the breakdown of higher-order structures causes the protein to **lose its biological activity**. Denatured proteins often precipitate, lose solubility, or fail to bind substrates properly, impairing physiological functions. Clinically, this phenomenon is highly relevant: for example, **high fever** in patients can lead to the denaturation of enzymes, resulting in slowed or impaired **metabolic processes**. In **parenteral (intravenous) nutrition**, improper pH or temperature of administered solutions can cause **plasma protein denaturation**, potentially affecting nutrient transport and immune function. Understanding the conditions that lead to protein denaturation is essential for nurses to ensure proper handling of medications, fluids, and patient care in fever, sepsis, or metabolic crises.

Clinical Examples

- **High fever** may denature enzymes like lactase or amylase, impairing digestion and energy metabolism.
- **IV nutrition solutions** with incorrect pH can denature serum proteins, altering fluid balance and drug efficacy.

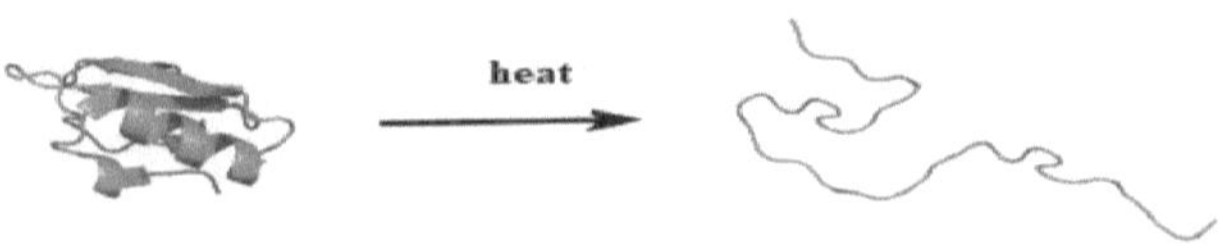

Figure 5.8 Protein Denaturation

Protein-Related Disorders, Malnutrition, and Clinical Monitoring

Mutations in the genetic code can alter the structure and function of proteins, leading to serious inherited diseases. A classic example is **sickle cell anemia**, in which a **point mutation** substitutes **glutamic acid with valine** in the β-globin chain of hemoglobin. This single amino acid change causes hemoglobin to misfold, leading to the distortion of red blood cells into a sickle shape, which impairs oxygen delivery and causes painful vaso-occlusive crises. Another significant disorder is **cystic fibrosis**, resulting from mutations in the

CFTR gene, which codes for a membrane protein that regulates chloride transport. Dysfunctional CFTR protein leads to the accumulation of **thick mucus** in the lungs and digestive tract, causing respiratory distress and malabsorption.

In clinical nutrition, **protein-energy malnutrition (PEM)** is a major concern, especially in children, the elderly, or hospitalized patients. **Kwashiorkor** occurs when protein intake is insufficient despite adequate caloric consumption. It presents with **edema**, **fatty liver**, and **poor wound healing**. In contrast, **marasmus** results from a severe deficiency in both protein and calories, leading to **muscle wasting**, **immune suppression**, and **developmental delays**. Nurses play a vital role in assessing protein status by monitoring **serum proteins** such as **albumin**, **prealbumin**, and **total protein levels**, especially in post-operative or critically ill patients.

From a diagnostic perspective, many proteins function as **biomarkers of disease**. For example, **troponin** is elevated in myocardial infarction, **amylase** in pancreatitis, and **ALT/AST** in liver disease. **Albumin**, a major plasma protein, is used to evaluate nutritional status, liver function, and fluid balance. Furthermore, many drugs bind to plasma proteins for distribution, and fluctuations in protein levels can affect **drug bioavailability, efficacy, and toxicity**, making **protein monitoring critical in pharmacologic care**.

Nursing Practice Applications

- Interpret lab results including total protein, albumin, globulins, and enzymes.
- Assess protein adequacy in patients on parenteral nutrition.
- Provide nutritional education, especially high-protein diets during recovery.
- Recognize and help manage autoimmune diseases, where antibodies target self-tissues.

Table 5.9 : Key Protein Disorders and Clinical Implications

Disorder / Condition	Protein Involved	Clinical Impact
Sickle Cell Anemia	Hemoglobin (β-chain)	Misshapen RBCs, poor oxygen delivery, pain crises
Cystic Fibrosis	CFTR (membrane channel)	Mucus buildup in lungs, pancreas, impaired digestion
Kwashiorkor	Dietary protein deficiency	Edema, delayed healing, fatty liver
Marasmus	Total energy/protein lack	Severe wasting, immune suppression, growth failure
Myocardial Infarction	Troponin	Cardiac enzyme elevated during heart damage
Liver Disease	ALT, AST, Albumin	Enzyme elevation and low albumin indicating hepatic dysfunction
Malnutrition / Protein Deficiency	Albumin, Prealbumin	Poor wound healing, edema, altered drug binding
Autoimmune Diseases	Immunoglobulins (antibodies)	Self-reactive antibodies causing tissue damage.

Enzymes: Function and Clinical Significance

Enzymes are specialized protein molecules that serve as biological catalysts—substances that accelerate chemical reactions without being altered or consumed in the process. In the human body, enzymes are indispensable for sustaining life, as nearly every metabolic process—ranging from digestion and nutrient absorption to DNA replication and energy production—relies on enzymatic activity. By lowering the activation energy required for reactions to proceed, enzymes ensure that essential biochemical transformations occur rapidly and efficiently under the mild conditions of body temperature and pH. Each enzyme is highly specific, acting on a particular substrate to produce a specific product, often regulated by the body's needs and environmental cues. For nurses and healthcare professionals, understanding the function and clinical significance of enzymes is essential for interpreting diagnostic tests, managing enzyme-related disorders, and administering treatments that depend on enzymatic pathways.

Function of Enzymes

Enzymes function by lowering the activation energy needed for chemical reactions, enabling these reactions to occur more rapidly and efficiently under the mild conditions of the human body, such as normal temperature and pH. Each enzyme exhibits a high degree of specificity for its substrate—a feature often illustrated by the "lock-and-key" model or the more flexible "induced fit" model. Most enzyme names end in "-ase" and are indicative of their function; for example:

- **Amylase** catalyzes the breakdown of starch into simple sugars
- **Lipase** digests fats
- **Lactase** breaks down lactose (milk sugar)
- **DNA polymerase** synthesizes DNA

To perform their catalytic roles effectively, many enzymes depend on **cofactors** such as metal ions (e.g., zinc, magnesium) or **coenzymes**, which are organic molecules often derived from vitamins (e.g., B-complex vitamins). These auxiliary components are critical for enzyme activation and stability, making proper nutrient intake important for maintaining enzymatic function and overall metabolic health.

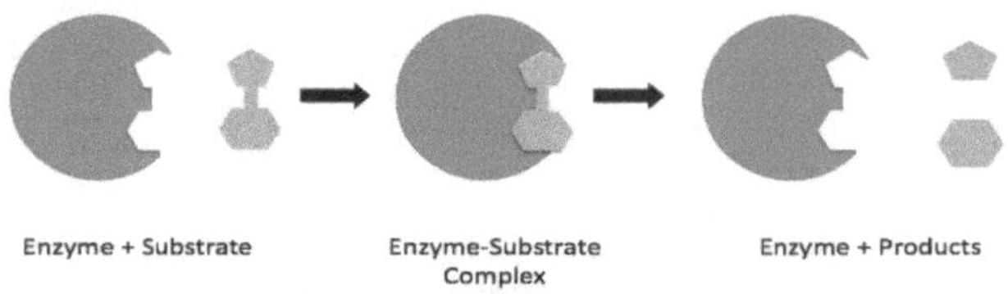

Figure 5.9 Mechanism of Enzyme Action

Clinical Significance of Enzymes

Beyond their fundamental role in metabolism, enzymes serve as crucial clinical biomarkers, as alterations in their levels or activity in blood or tissues often signal underlying pathological conditions. For example, cardiac enzymes such as troponin and creatine kinase-MB (CK-MB) are routinely measured in suspected cases of myocardial infarction, where elevated levels indicate cardiac muscle injury.

Liver function tests depend on enzymes like ALT (alanine aminotransferase), AST (aspartate aminotransferase), ALP (alkaline phosphatase), and GGT (gamma-glutamyl transferase) to identify hepatocellular damage, cholestasis, or inflammation as seen in hepatitis. Similarly, elevated levels of pancreatic enzymes, such as amylase and lipase, are indicative of acute pancreatitis. Another example is lactate dehydrogenase (LDH), a general marker of tissue damage that can occur in the heart, liver, kidneys, muscles, and other organs. Monitoring enzyme levels is thus vital in diagnosing diseases, evaluating their severity, and assessing the effectiveness of medical treatment.

Table 5.10 : Clinically Important Enzymes and Their Diagnostic Significance

Enzyme	Primary Source	Clinical Use / Diagnostic Significance
Troponin I / T	Cardiac muscle	Highly specific marker for myocardial infarction (heart attack)
Creatine Kinase-MB (CK-MB)	Heart muscle	Elevated in myocardial infarction and cardiac muscle injury
ALT (Alanine Aminotransferase)	Liver	Indicates liver cell damage; elevated in hepatitis
AST (Aspartate Aminotransferase)	Liver, heart	Elevated in liver disease, myocardial infarction, or muscle injury
ALP (Alkaline Phosphatase)	Liver, bones	Elevated in bile duct obstruction, liver disease, and bone disorders
GGT (Gamma-glutamyl transferase)	Liver, biliary tract	Marker of bile duct obstruction; elevated in alcoholic liver disease
Amylase	Pancreas, salivary glands	Elevated in acute pancreatitis and salivary gland disorders
Lipase	Pancreas	More specific than amylase for diagnosing acute pancreatitis
LDH (Lactate Dehydrogenase)	Various tissues	Indicates general tissue damage (e.g., heart, liver, muscle, cancer)

Enzyme Deficiencies and Associated Disorders

Enzyme deficiencies can lead to a wide spectrum of metabolic and physiological disorders, many of which carry serious clinical implications. A common example is lactase deficiency, which impairs the ability to digest lactose, the sugar found in milk and dairy products. This condition, known as lactose intolerance, results in symptoms such as bloating, abdominal pain, gas, and diarrhea after dairy consumption.

Phenylketonuria (PKU), caused by a deficiency of the enzyme phenylalanine hydroxylase, leads to toxic buildup of phenylalanine in the brain, causing intellectual disability, developmental delays, and neurological complications.

Glucose-6-phosphate dehydrogenase (G6PD) deficiency is an X-linked recessive disorder affecting red blood cell metabolism. Affected individuals are vulnerable to oxidative stress, which can trigger hemolytic anemia after exposure to certain medications, infections, or consumption of fava beans.

Understanding these enzyme-related conditions is crucial for early diagnosis, patient education, and nursing care planning.

Nursing Care for Enzyme Deficiency Disorders

Effective nursing care for enzyme deficiencies focuses on patient education and preventive strategies. Nurses should guide lactose-intolerant patients to avoid lactose-containing foods or use suitable alternatives. For phenylketonuria (PKU), supporting strict adherence to a low-phenylalanine diet and regular monitoring is essential. In cases of G6PD deficiency, educating patients to avoid known triggers such as specific medications and foods, while closely observing for signs of hemolytic anemia, is critical. Through comprehensive education and vigilant monitoring, nurses play a vital role in preventing complications and enhancing patient quality of life.

Table 5.11: Common Enzyme Deficiencies and Their Clinical Relevance

Enzyme Deficiency	Cause / Inheritance	Clinical Features	Nursing Considerations
Lactase Deficiency	Genetic or acquired with age	Bloating, gas, diarrhea after dairy	Lactose-free diet, monitor calcium/Vit D
Phenylalanine Hydroxylase Deficiency (PKU)	Autosomal recessive	Intellectual disability, seizures	Newborn screening, low-phenylalanine diet
G6PD Deficiency	X-linked recessive (males mostly)	Hemolytic anemia, jaundice	Avoid oxidative drugs, monitor infections

Therapeutic and Pharmaceutical Uses of Enzymes

Enzymes are not only central to human physiology but also have **critical applications in modern pharmacology and therapy**. In clinical practice, **digestive enzymes** such as **lipase, amylase,** and **protease** are incorporated into oral enzyme replacement therapies to aid patients with **pancreatic insufficiency**, enhancing **nutrient absorption** and alleviating **gastrointestinal symptoms**.

Enzymes also serve as **therapeutic targets**. A key example includes **enzyme inhibitors**, an important class of drugs. For instance, **angiotensin-converting enzyme (ACE) inhibitors** are widely used in managing **hypertension** and **heart failure** by **inhibiting the formation of angiotensin II**, a potent vasoconstrictor. Likewise, **protease inhibitors** are essential components of **HIV treatment**, where they block viral replication by inhibiting viral enzymes.

In genetic disorders, **recombinant enzyme therapies** have revolutionized treatment. For example, **enzyme replacement therapy (ERT)** in **Gaucher's disease** uses **synthetic glucocerebrosidase** to compensate for deficient enzyme activity, preventing organ damage and improving patient outcomes.

These examples demonstrate how **enzyme science directly translates into therapeutic advances**, enabling targeted treatments and personalized care.

The Critical Role of Enzymes in Nursing Assessment and Care

For nurses, a thorough understanding of enzyme activity is vital for accurate diagnosis and optimal patient management. Enzymes frequently act as key biomarkers in clinical practice—for instance, troponin indicates myocardial infarction, ALT and AST reflect liver function, and amylase and lipase are markers for pancreatic health. Therefore, nurses must skillfully interpret and monitor these laboratory values within the broader clinical context.

Recognizing symptoms associated with enzyme deficiencies, such as digestive disturbances in lactase deficiency or acute hemolysis in G6PD deficiency, enables prompt intervention and improved patient outcomes. Nurses also play an essential role in educating patients about necessary lifestyle and dietary adjustments, including lactose-free diets or phenylalanine-restricted regimens.

Furthermore, since many medications interact with enzymatic pathways, nurses should be knowledgeable about enzyme-related drug mechanisms, vigilant for adverse effects, and capable of identifying early signs of complications. It is important to evaluate enzyme levels over time rather than in isolation to detect meaningful clinical trends. In emergencies, such as chest pain, timely interpretation of cardiac enzyme levels can be lifesaving.

Lipids

Lipids are a diverse group of hydrophobic or amphipathic biomolecules primarily composed of carbon (C), hydrogen (H), and oxygen (O), though they contain significantly less oxygen than carbohydrates. Due to their predominantly nonpolar nature, lipids are insoluble in water but readily dissolve in organic solvents such as alcohol, ether, and chloroform. In human biology, lipids play vital roles as concentrated energy reserves, structural components of cellular membranes, and precursors for bioactive molecules such as steroid hormones and prostaglandins. Additionally, lipids serve as thermal insulators, cushioning organs and helping regulate body temperature. Their structural diversity—ranging from triglycerides and phospholipids to steroids and essential fatty acids—enables a broad spectrum of physiological functions that are critical to health and homeostasis.

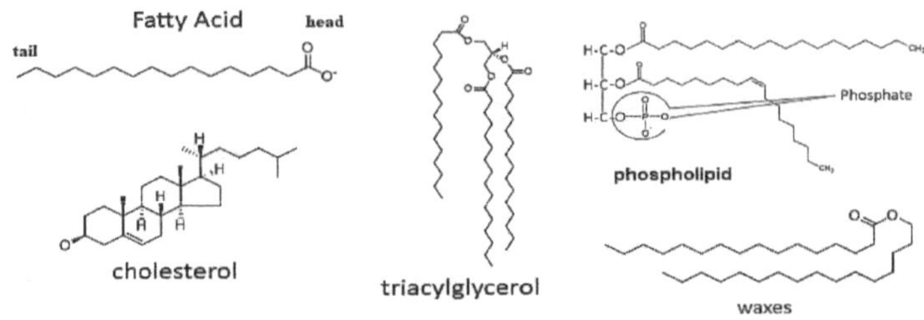

Figure 5.10 Types of lipids

Major Classes of Lipids and Their Roles in the Body

Lipids are structurally and functionally diverse molecules that are classified into several major categories based on their chemical structure and physiological roles. A clear understanding of these lipid classes is essential in clinical practice, as lipid imbalances are implicated in numerous chronic conditions, including cardiovascular disease, obesity, and metabolic syndrome.

Fatty acids are the simplest lipids, consisting of long hydrocarbon chains (typically 12–24 carbon atoms) ending with a carboxyl group (–COOH). They are broadly classified into **saturated** and **unsaturated** fatty acids. Saturated fatty acids, which contain no carbon-carbon double bonds, are usually solid at room temperature and found in animal fats, dairy products, and tropical oils like coconut oil. High intake of saturated fats is linked to increased LDL ("bad") cholesterol and a heightened risk of cardiovascular disease. In contrast, unsaturated fatty acids have one or more double bonds and are typically liquid at room temperature. These are subdivided into **monounsaturated fats** (e.g., oleic acid in olive oil) and **polyunsaturated fats** (e.g., linoleic and α-linolenic acids found in plant oils, nuts, and fish). Among them, **essential fatty acids (EFAs)** like omega-3 and omega-6 must be obtained from the diet. They play vital roles in reducing inflammation, supporting brain development, maintaining membrane fluidity, and promoting cardiovascular health.

Triglycerides, also known as neutral fats, are the most abundant lipids in the body and the main form of long-term energy storage. Each triglyceride is composed of three fatty acid chains esterified to a glycerol backbone. Stored primarily in adipose tissue, triglycerides are mobilized during fasting or physical exertion to meet energy needs. Because they provide over twice as much energy per gram as carbohydrates or proteins, they serve as an efficient fuel reserve. Additionally, triglycerides offer insulation and cushion internal organs. Clinically, elevated triglyceride levels are associated with obesity, cardiovascular disease, and non-alcoholic fatty liver disease (NAFLD), making them important biomarkers in metabolic screening.

Phospholipids are amphipathic molecules composed of a glycerol backbone, two fatty acid tails, and a phosphate-containing polar head. They are essential structural components of cell membranes, forming the phospholipid bilayer that controls the movement of substances in and out of cells. Beyond their structural role, phospholipids participate in nerve conduction, signal transduction, and the formation of pulmonary surfactant. In clinical contexts, phospholipid deficiency—especially in premature infants—can lead to **respiratory distress syndrome (RDS)** due to inadequate surfactant production.

Steroids are a distinct class of lipids characterized by a core structure of four fused carbon rings. **Cholesterol** is the most prominent steroid, serving as a precursor to steroid hormones (e.g., cortisol, aldosterone), sex hormones (e.g., estrogen, testosterone), vitamin D, and bile salts. It also contributes to membrane integrity and fluidity. Elevated LDL cholesterol levels are a major risk factor for atherosclerosis and heart disease. Statins, widely used lipid-lowering drugs, act by inhibiting cholesterol synthesis in the liver.

Waxes and lipoproteins form the final medically relevant lipid categories. **Waxes**, composed of long-chain fatty acids and alcohols, function as protective coatings in plants and animals. In humans, they are found in skin secretions and cerumen (earwax), providing waterproofing and antimicrobial defense. **Lipoproteins** are complexes of lipids and proteins that transport hydrophobic lipids through the aqueous bloodstream. They are categorized by density: **high-density lipoproteins (HDL)** remove cholesterol from tissues and are considered "good cholesterol," while **low-density lipoproteins (LDL)** deposit cholesterol in arteries and are labeled "bad cholesterol." **Very-low-density lipoproteins (VLDL)** transport triglycerides from the liver to peripheral tissues. Monitoring lipoprotein levels is essential in assessing cardiovascular risk and guiding treatment strategies.

Quick Nursing Tip

It is essential for nurses to closely monitor lipid profiles—including total cholesterol, HDL, LDL, and triglycerides—in patients who have cardiovascular risk factors such as hypertension, diabetes, obesity, or a family history of heart disease. Regular assessment of these lipid levels helps in early identification of dyslipidemia, enabling timely interventions to prevent progression of cardiovascular disease. Nurses should also provide comprehensive dietary guidance that encourages the intake of healthy unsaturated fats found in fish, nuts, seeds, and vegetable oils while advising patients to reduce consumption of saturated and trans fats commonly present in processed and fried foods. Promoting these dietary modifications supports the reduction of atherosclerosis risk and helps manage cholesterol levels effectively. Additionally, educating patients on lifestyle changes, including regular physical activity and smoking cessation, further enhances cardiovascular health and improves overall patient outcomes.

Table 5.12: Major Types of Lipids

Type of Lipid	Structure	Biological Function	Clinical Significance
Fatty Acids	Long hydrocarbon chain + –COOH	Energy source; lipid building blocks	Saturated fats raise LDL; EFAs vital for heart & brain
Triglycerides	3 fatty acids + glycerol	Long-term energy storage; insulation	Linked to obesity, heart disease
Phospholipids	2 fatty acids + glycerol + phosphate	Cell membrane component; amphipathic	Deficiency causes respiratory issues; cell signaling
Steroids	Four fused carbon rings	Hormone, bile salt, vitamin D precursors	High cholesterol → atherosclerosis; hormone precursors
Waxes & Lipoproteins	Waxes: esters; Lipoproteins: lipid-protein	Waterproof barrier; lipid transport	HDL good, LDL bad cholesterol

Biological Roles and Clinical Relevance of Lipids

Lipids are not merely energy-dense molecules; they perform a wide array of structural, regulatory, and protective functions essential for maintaining human health. Their structural diversity—including fatty acids, triglycerides, phospholipids, steroids, and lipoproteins—enables them to participate in numerous physiological processes.

Energy Storage and Mobilization :Triglycerides, stored in adipose tissue, are the body's principal energy reservoir, yielding approximately 9 kcal per gram—more than twice the energy provided by carbohydrates or proteins. During fasting or extended physical activity, fatty acids are mobilized and oxidized in mitochondria to produce ATP, sustaining metabolic demands.

Membrane Structure and Integrity: Phospholipids and cholesterol form the lipid bilayer of cell membranes, a dynamic structure that provides membrane fluidity, mechanical stability, and selective permeability. This architecture is crucial for nutrient transport, signal transduction, and intercellular communication.

Thermal and Physical Protection: Adipose tissue insulates the body, conserves heat, and cushions vital organs such as the kidneys and eyes. Similarly, the lipid-rich myelin sheath surrounding neurons enhances electrical conduction and protects nerve fibers.

Hormone and Vitamin Precursors: Cholesterol serves as the biochemical precursor for steroid hormones (e.g., cortisol, aldosterone, estrogen, testosterone), vitamin D, and bile acids that facilitate lipid digestion. Essential fatty acids (omega-3 and omega-6) give rise to prostaglandins and leukotrienes, bioactive lipids that modulate inflammation, immunity, and blood clotting.

Clinical Implications: Dysregulation of lipid metabolism is implicated in several major health conditions. Elevated LDL cholesterol and triglyceride levels contribute to atherosclerosis, increasing the risk of myocardial infarction and stroke. In contrast, HDL cholesterol exerts a protective effect by facilitating reverse cholesterol transport.

Nurses play a pivotal role in managing lipid-related conditions by:

- Monitoring lipid profiles
- Educating patients on heart-healthy diets
- Supporting adherence to pharmacological interventions like statins

Obesity and Metabolic Syndrome: Excess triglyceride storage underlies central obesity and is associated with insulin resistance, hypertension, and abnormal lipid profiles—hallmarks of metabolic syndrome. Nursing interventions such as lifestyle counseling, weight management programs, and nutritional education are crucial in combating these conditions.

Malabsorption and Lipid Deficiency: Diseases such as celiac disease, Crohn's disease, and bile salt deficiency can impair lipid absorption, leading to deficiencies in fat-soluble vitamins (A, D, E, and K) and essential fatty acids. Clinical signs include xerosis (dry skin), night blindness, bleeding disorders, and weakened immune function.

Neurological and Neonatal Considerations: Lipid abnormalities affect the nervous system. In multiple sclerosis, degradation of the myelin sheath impairs neuronal function. Lipid metabolism disturbances have also been linked to neurodegenerative conditions like Alzheimer's disease. In neonatology, premature infants may suffer from respiratory distress syndrome (RDS) due to insufficient surfactant—a phospholipid-based compound essential for lung function.

Nursing Responsibilities: Nurses must be adept at interpreting lipid panels and understanding their implications for cardiovascular, hepatic, and metabolic health. In patients receiving parenteral nutrition, lipid supplementation is critical to meet energy needs and maintain cellular integrity. Vigilance for signs of lipid excess or deficiency is especially important in critical care settings, where metabolic demands are heightened.

Nucleic Acids: DNA and RNA Structure, and Their Role in Protein Synthesis and Heredity

Nucleic acids are essential macromolecules that store, transmit, and regulate genetic information in all living organisms. There are two primary types: **DNA (deoxyribonucleic acid)** and **RNA (ribonucleic acid)**. DNA carries the permanent genetic blueprint of the cell, while RNA plays a central role in decoding and executing that information during protein synthesis.

DNA Structure and Function: DNA consists of two long strands arranged in a **double helix**, with each strand made of repeating units called **nucleotides**. Each nucleotide comprises a **deoxyribose sugar**, a **phosphate group**, and one of four **nitrogenous bases**: **adenine (A), thymine (T), cytosine (C),** or **guanine (G)**. The strands are held together by **hydrogen bonds** between complementary bases: A pairs with T, and C pairs with G. This base pairing ensures accurate DNA replication, stability, and faithful inheritance of genetic material.

RNA Structure and Function: RNA is generally **single-stranded** and differs from DNA by containing **ribose sugar** instead of deoxyribose, and **uracil (U)** in place of thymine. RNA molecules are crucial in **gene expression**, particularly in **protein synthesis**. There are three main types of RNA:

- **Messenger RNA (mRNA):** carries genetic instructions from DNA in the nucleus to ribosomes in the cytoplasm.
- **Transfer RNA (tRNA):** transports specific amino acids to the ribosome based on the mRNA code.
- **Ribosomal RNA (rRNA):** forms the structural and enzymatic core of ribosomes and catalyzes peptide bond formation.

Genetic Expression and Heredity: Together, DNA and RNA regulate **gene expression**, determining which proteins are produced in each cell and, ultimately, governing cellular structure and function. DNA also ensures **heredity**, as it is accurately duplicated and passed from one generation to the next during **cell division and reproduction**. Mutations in DNA or RNA can disrupt normal gene function and lead to **genetic disorders**, **cancers**, or **inherited diseases**. Therefore, a thorough understanding of nucleic acid structure and function is critical in the fields of **medicine, genetics**, and **nursing practice**.

Metabolism Overview

Metabolism encompasses all chemical reactions occurring within living cells that are essential for sustaining life. These processes convert nutrients into usable energy, build essential biomolecules, and eliminate waste products. Metabolism is broadly categorized into two interdependent processes:

- **Catabolism:** the breakdown of complex molecules (e.g., carbohydrates, lipids, proteins) into simpler units, releasing energy primarily in the form of ATP.
- **Anabolism:** the synthesis of complex molecules (e.g., proteins, DNA, cell membranes) from simpler ones, requiring energy input derived from ATP.

These two processes are intricately linked and operate in a dynamic balance to support cellular function and organismal health. Catabolic pathways provide the necessary energy and building blocks for anabolic pathways, while anabolic pathways repair and replace damaged cellular components, enable growth, and support reproduction. The integration of these pathways ensures that energy production and biosynthesis are coordinated to meet the body's fluctuating demands. For instance, during periods of increased physical activity or stress, catabolism is upregulated to provide extra energy, whereas during rest or growth phases, anabolism predominates to rebuild tissues and store energy.

The primary goal of metabolism is to produce sufficient energy to sustain vital cellular activities while supplying the body with the necessary molecules to construct and repair tissues. Key metabolic pathways involving carbohydrates, fats, and proteins are tightly regulated by enzymes and hormones such as insulin and glucagon, which respond to nutrient availability and physiological needs. ATP (adenosine triphosphate) serves as the universal energy currency, transferring energy from catabolic reactions to power anabolic processes and other cellular functions.

Maintaining a proper balance between catabolism and anabolism is critical for health. Disruption of this balance can lead to clinical conditions such as cachexia, characterized by excessive catabolism and muscle wasting, or obesity, resulting from excessive anabolism and energy storage.

For nurses, a thorough understanding of metabolism is essential for effective patient care. It aids in managing diseases like diabetes, thyroid disorders, malnutrition, obesity, and acid–base imbalances. Moreover, it helps in interpreting laboratory results, guiding nutritional interventions, and understanding pharmacological treatments that influence metabolic pathways. Recognizing how these processes interact empowers nurses to support patients' recovery and maintain metabolic homeostasis.

Table 5.13: Comparison Between Catabolism and Anabolism

Aspect	Catabolism	Anabolism
Definition	Breakdown of complex molecules into simpler ones	Synthesis of complex molecules from simpler ones
Energy Flow	Releases energy (exergonic) — produces ATP	Requires energy input (endergonic) — consumes ATP
Purpose	Provides energy and building blocks for the body	Supports growth, repair, and storage
Examples	Glucose breakdown (glycolysis), lipid oxidation, protein degradation	Protein synthesis, DNA replication, cell membrane formation
ATP Role	ATP is generated	ATP is consumed
Associated Conditions	Excess catabolism → cachexia (e.g., in chronic illness or cancer)	Excess anabolism → obesity, excessive tissue or fat buildup
Hormonal Regulation	Stimulated by glucagon, adrenaline, cortisol	Stimulated by insulin, growth hormone, sex hormones

ATP – The Energy Currency of the Cell

Adenosine triphosphate (ATP) serves as the universal energy currency in all living organisms, driving virtually every energy-requiring process within the cell. Structurally, ATP is composed of an adenine nitrogenous base, a five-carbon ribose sugar, and three phosphate groups linked by high-energy phosphoanhydride bonds. The most critical of these is the bond between the second and third phosphate groups, which stores a significant amount of potential energy. When a cell requires energy to perform work, ATP undergoes hydrolysis—a chemical reaction in which the terminal phosphate bond is broken—yielding adenosine diphosphate (ADP) and an inorganic phosphate (Pi). This reaction releases energy that is immediately harnessed to fuel essential cellular activities, including muscle contraction, nerve impulse transmission, active transport of molecules across cell membranes, and the biosynthesis of macromolecules such as proteins and nucleic acids.

Because ATP cannot be stored in large quantities, it must be regenerated continuously through catabolic pathways that extract energy from nutrients. These pathways include glycolysis (the anaerobic breakdown of glucose in the cytoplasm), the citric acid cycle (Krebs cycle) in the mitochondria, and oxidative phosphorylation, where the majority of ATP is produced via the electron transport chain. These processes ensure that cells maintain a steady supply of ATP to meet both routine and high-demand energy needs.

For nurses, understanding the dynamics of ATP production and consumption is essential for evaluating the metabolic status of patients—especially those who are critically ill, hypoxic, or suffering from metabolic disorders. It also provides the biochemical rationale behind interventions such as IV fluid composition (e.g., glucose and electrolyte content), electrolyte therapy to support cellular energetics, and tailored nutritional support to sustain ATP production and promote recovery. In clinical settings, an awareness of how energy is generated and utilized at the molecular level is crucial for optimizing patient care and ensuring effective physiological support.

Carbohydrate metabolism

Carbohydrate metabolism refers to the series of biochemical processes through which the body converts carbohydrates from food into usable energy. This process is essential for maintaining cellular activity, brain function, and overall energy balance.

The process begins in the **mouth**, where **salivary amylase** initiates the breakdown of complex carbohydrates (starches) into simpler sugars. Digestion continues in the **small intestine**, where enzymes further convert carbohydrates into **glucose**—the body's primary and most efficient energy source.

Glucose is absorbed into the **bloodstream**, raising **blood sugar levels**. In response, the **pancreas** releases **insulin**, a hormone that facilitates glucose uptake by body cells. Once inside the cells, glucose enters the **glycolysis** pathway—a sequence of enzymatic reactions in the cytoplasm that convert glucose into **pyruvic acid**, yielding a small amount of **ATP**.

Pyruvic acid then enters the **mitochondria**, where it undergoes complete oxidation through the **Krebs cycle** and the **electron transport chain**. These processes produce a significant amount of **ATP**, along with **carbon dioxide (CO_2)** and **water (H_2O)** as byproducts.

If glucose is abundant, the body stores it as **glycogen** in the **liver and skeletal muscles**. When glycogen stores are full, excess glucose is converted into **fat** for long-term energy storage.

Carbohydrate metabolism is crucial for fueling muscle activity, supporting brain function (which depends almost exclusively on glucose), and maintaining metabolic homeostasis.

Glycolysis: The First Step in Energy Harvest

Glycolysis is a fundamental metabolic pathway that initiates the cellular breakdown of **glucose**—a six-carbon sugar—into **two molecules of pyruvate**, each containing three carbon atoms. This process takes place in the **cytoplasm** of all cells and occurs **anaerobically** (without the need for oxygen), making it essential for cells that lack mitochondria, such as **red blood cells**.

Glycolysis consists of **ten enzyme-catalyzed steps** and results in a **net gain of 2 ATP molecules** and **2 NADH molecules** per glucose molecule. The ATP provides immediate energy for cellular activities, while NADH acts as an electron carrier, potentially generating more ATP through later aerobic processes if oxygen is available.

This pathway plays a critical role in conditions of **oxygen deficiency** (e.g., tissue hypoxia, ischemia, or vigorous muscle activity). When oxygen is available (**aerobic conditions**), **pyruvate** enters the mitochondria and continues through the **Krebs cycle** and **electron transport chain** for full energy extraction. Under **anaerobic**

conditions, however, pyruvate is converted into **lactate**. This reaction regenerates **NAD**$^+$, allowing glycolysis to continue despite the lack of oxygen.

For nursing professionals, understanding glycolysis is vital in:

- Interpreting **lactic acidosis**
- Managing patients with **hypoxia**
- Recognizing how **metabolic disorders** or certain **medications** affect cellular energy production and overall tissue function

Pyruvate Oxidation: The Gateway to Aerobic Metabolism

Following **glycolysis**, if oxygen is available, each **pyruvate** molecule is transported into the **mitochondrial matrix**, where it undergoes **pyruvate oxidation**. This crucial transition is catalyzed by the **pyruvate dehydrogenase complex (PDC)**, which:

- Removes one carbon from pyruvate as **carbon dioxide (CO_2)**
- Converts the remaining two-carbon fragment into an **acetyl group**
- Links the acetyl group to **coenzyme A (CoA)**, forming **acetyl-CoA**
- Generates **one molecule of NADH**, which carries high-energy electrons to the electron transport chain

Pyruvate oxidation forms a metabolic bridge between **anaerobic glycolysis** and the **aerobic Krebs cycle**. This step is tightly regulated by the cell's energy status: high levels of **ATP**, **NADH**, or **acetyl-CoA** inhibit the process, conserving resources when energy demand is low.

For nurses, understanding pyruvate oxidation is critical in recognizing how **oxygen availability** affects energy metabolism. Clinical conditions such as **hypoxia**, **sepsis**, or **mitochondrial disorders** may impair this step, leading to **lactate accumulation** and **metabolic acidosis**, indicating a shift toward less efficient **anaerobic metabolism**.

Krebs Cycle: Central Hub of Cellular Respiration

Once **acetyl-CoA** is formed, it enters the **Krebs cycle** (also known as the **citric acid cycle** or **TCA cycle**), a core aerobic pathway occurring in the **mitochondrial matrix**. This cycle completes the oxidation of carbon from glucose, extracting energy through a series of enzyme-driven reactions.

For **each turn** of the cycle, the following products are generated:

- **3 NADH** molecules
- **1 $FADH_2$** molecule
- **1 GTP**, which is easily converted into **ATP**
- **2 CO_2** molecules as waste

NADH and **$FADH_2$** are essential electron carriers that transport high-energy electrons to the **electron transport chain**, where the bulk of **ATP** is synthesized via **oxidative phosphorylation**.

The Krebs cycle operates **only under aerobic conditions** and is regulated through feedback mechanisms based on energy availability and substrate levels.

Clinically, disruption of the Krebs cycle—due to **hypoxia, vitamin B deficiencies** (e.g., **niacin**, **riboflavin**), or **mitochondrial dysfunction**—can significantly impair **ATP production**. For nursing professionals, knowledge of this cycle aids in understanding conditions like **metabolic acidosis, energy failure in critical illness**, and the rationale for **nutritional and pharmacologic interventions** targeting cellular energy restoration.

Electron Transport Chain and Oxidative Phosphorylation: The Final Step in Energy Harvest

The electron transport chain (ETC), embedded in the inner mitochondrial membrane, is the culminating stage of aerobic respiration. It receives high-energy electrons from NADH and $FADH_2$—produced during glycolysis, pyruvate oxidation, and the Krebs cycle—and passes them through a series of protein complexes (I–IV). As electrons flow down this chain, energy is released and used to actively pump protons (H^+) from the mitochondrial matrix into the intermembrane space, creating an electrochemical gradient known as the proton motive force.

This gradient represents stored potential energy. Protons then flow back into the matrix through a specialized enzyme, ATP synthase, which harnesses this energy to phosphorylate ADP into ATP. This process—known as oxidative phosphorylation—can generate up to 34 ATP molecules per glucose molecule, making it the most efficient step in cellular respiration.

Oxygen serves as the final electron acceptor in the chain, combining with electrons and protons to form water. Without oxygen, the ETC halts, and ATP synthesis drops dramatically, forcing cells to rely on less efficient anaerobic metabolism.

From a clinical perspective, the ETC is essential in high-energy-demand tissues like the brain and heart. Its disruption—whether due to oxygen deprivation (hypoxia), mitochondrial toxins (e.g., cyanide), or genetic mitochondrial disorders—can lead to life-threatening energy failure. For nurses, understanding this pathway is crucial in managing critically ill patients, recognizing signs of mitochondrial dysfunction, and appreciating how oxygen therapy or metabolic support may restore ATP production.

ATP Production: Fueling the Functions of Life

The ultimate purpose of cellular metabolism is the efficient production of adenosine triphosphate (ATP), the universal energy currency of the cell. ATP powers nearly every biological process, including muscle contraction, nerve impulse transmission, biosynthesis of macromolecules, active transport of ions, and maintenance of cellular homeostasis. Through the coordinated sequence of glycolysis, pyruvate oxidation, the Krebs cycle, and oxidative phosphorylation, a single molecule of glucose can yield up to 36–38 molecules of ATP under optimal aerobic conditions.

Each metabolic stage contributes to this energy yield: glycolysis provides 2 ATP and 2 NADH; pyruvate oxidation and the Krebs cycle supply more NADH and $FADH_2$; and the electron transport chain, using these carriers, drives the bulk of ATP generation via oxidative phosphorylation. The efficiency of this system depends heavily on the availability of oxygen and the functional integrity of mitochondria.

Clinically, ATP depletion is a hallmark of critical conditions such as ischemia, sepsis, and mitochondrial disorders. Tissues with high metabolic demands—like cardiac muscle, neurons, and renal tubules—are especially vulnerable to ATP deficits, which can lead to dysfunction or cell death. For nurses, understanding ATP dynamics is essential in assessing patient energy status, supporting nutritional and oxygen therapy, and recognizing early signs of metabolic crisis.

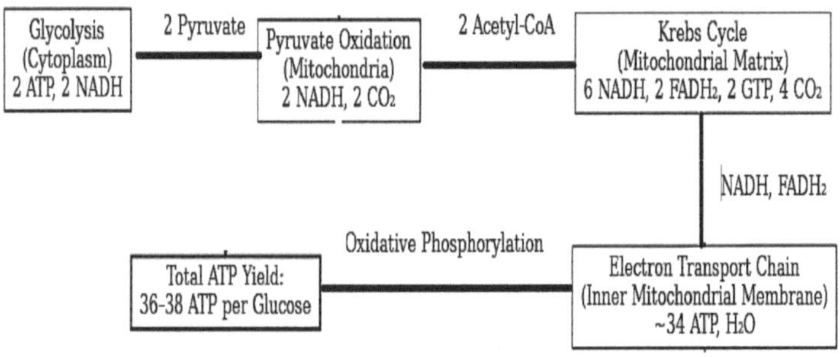

Figure 5.10 Metabolic Pathway of ATP Production from Glucose

Protein Metabolism: Building and Balancing the Body's Functional Framework

Protein metabolism encompasses the intricate biochemical processes by which dietary proteins are digested, absorbed, utilized, and regulated within the body. It begins in the stomach, where hydrochloric acid (HCl) and the enzyme pepsin work together to denature and partially hydrolyze proteins into shorter polypeptides. These peptides are further digested in the small intestine by enzymes such as trypsin, chymotrypsin, and peptidases, producing free amino acids that are readily absorbed into the bloodstream.

Once in circulation, amino acids are distributed to tissues and utilized primarily for anabolic functions, including:

- Synthesis of structural and contractile proteins (e.g., in muscles and connective tissue)
- Production of enzymes, hormones, and neurotransmitters
- Immune support via antibody and cytokine synthesis
- Wound healing and tissue repair

Unlike carbohydrates and fats, proteins are not the body's preferred energy source. However, during prolonged fasting, severe illness, or intense physical activity, amino acids may be diverted for energy through gluconeogenesis or oxidized via the Krebs cycle.

Importantly, the body cannot store excess amino acids. Surplus amino acids undergo deamination in the liver, a process that removes the nitrogen-containing amino group. This nitrogen is converted into urea and excreted by the kidneys. The remaining carbon skeleton may then be used for energy or transformed into glucose or fatty acids for storage.

For healthcare professionals, understanding protein metabolism is vital in managing nutritional therapy, renal and liver function, catabolic states, and protein-energy malnutrition. It is a cornerstone in assessing the body's recovery needs, especially in critically ill, post-surgical, or malnourished patients.

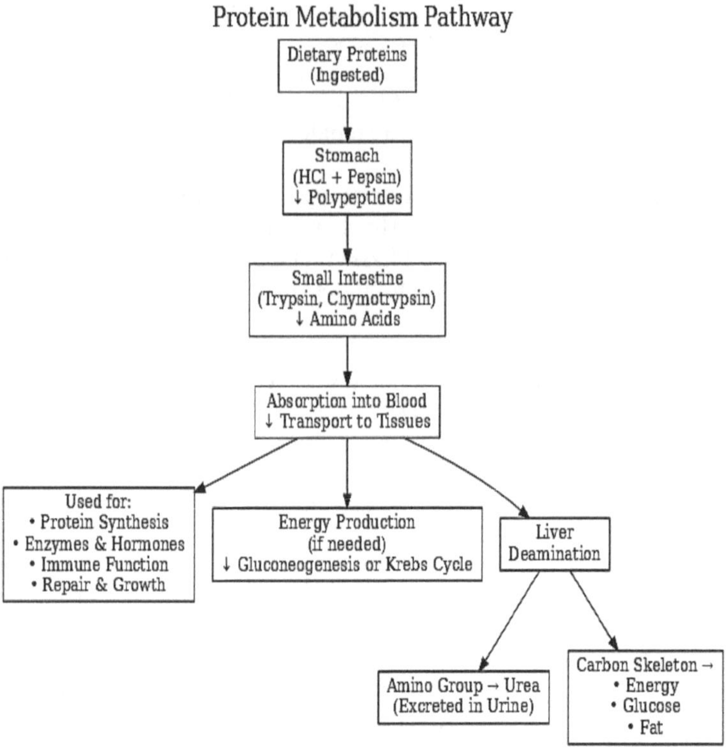

Figure 5.11 protein Metabolic Pathway

Lipid Metabolism: Harnessing Fats for Energy and Storage

Lipid metabolism refers to the series of physiological processes through which the body digests, absorbs, transports, utilizes, and stores fats to maintain **energy balance** and support **cellular function**. This metabolic pathway is essential for **energy production**, particularly during periods of **fasting**, **physical activity**, or **low carbohydrate availability**.

1. Digestion and Absorption

Lipid metabolism begins in the **small intestine**, where dietary fats—mainly **triglycerides**—are emulsified by **bile salts** secreted by the

liver and stored in the **gallbladder**. Emulsification breaks large lipid globules into smaller **micelles**, increasing surface area for digestion.

Pancreatic lipase then hydrolyzes triglycerides into **free fatty acids** and **glycerol**. These products are absorbed by **intestinal epithelial cells (enterocytes)**, where they are reassembled into triglycerides and packaged into **chylomicrons**—lipoprotein particles that transport lipids via the **lymphatic system** and eventually into **systemic circulation**.

2. Transport and Utilization

Chylomicrons deliver lipids to **skeletal muscle, cardiac muscle**, and **adipose tissue**. At target tissues, **lipoprotein lipase** breaks triglycerides down again into free fatty acids and glycerol. Fatty acids enter cells and are transported into **mitochondria**, where they undergo **beta-oxidation**—a process that sequentially removes two-carbon units, forming **acetyl-CoA**.

Acetyl-CoA enters the **Krebs cycle**, producing **NADH** and **$FADH_2$**, which transfer high-energy electrons to the **electron transport chain** for **ATP synthesis** via **oxidative phosphorylation**. This yields a **high amount of ATP**, making fat metabolism highly efficient in energy production.

3. Storage and Mobilization

When **energy intake exceeds demand**, excess fatty acids are stored in **adipose tissue** as triglycerides. During **fasting, exercise**, or **carbohydrate restriction**, these triglycerides are broken down by **hormone-sensitive lipases**, releasing fatty acids into the bloodstream for ATP generation.

4. Ketone Body Formation

When **acetyl-CoA accumulates** beyond the capacity of the Krebs cycle—such as in **starvation**, **prolonged fasting**, or **uncontrolled diabetes**—the **liver** converts it into **ketone bodies** (acetoacetate, β-hydroxybutyrate, and acetone). These molecules serve as an **alternative energy source**, especially for the **brain**, **heart**, and **muscles** when glucose is scarce.

For nurses, understanding lipid metabolism is vital for delivering effective, evidence-based care in a variety of clinical settings. It plays a central role in managing nutritional therapy, including the implementation and monitoring of ketogenic diets. Nurses must also be proficient in interpreting lipid panel results to assess cardiovascular risk and guide patient education. In addition, knowledge of lipid pathways is essential when addressing metabolic disorders such as hyperlipidemia, ketosis, and non-alcoholic fatty liver disease (NAFLD). During periods of fasting, in the management of diabetes, or in critically ill patients, nurses rely on their understanding of lipid metabolism to support energy balance, tailor nutritional interventions, and monitor for signs of metabolic dysfunction.

Case Study: Lipid Metabolism in a Diabetic Patient with NAFLD

Mr. Nasser, a 58-year-old male with a history of type 2 diabetes and obesity (BMI 34), was recently diagnosed with non-alcoholic fatty liver disease (NAFLD). He presented with complaints of fatigue and poor glycemic control. Laboratory tests revealed elevated triglycerides (245 mg/dL), high LDL cholesterol (180 mg/dL), low HDL (32 mg/dL), and mildly increased liver enzymes. His HbA1c was 8.5%, indicating poorly controlled diabetes.

The medical team recommended a low-carbohydrate, moderate-fat dietary intervention to promote weight loss and reduce hepatic fat accumulation. As the body adapts to reduced glucose intake, it shifts toward lipid metabolism for energy, increasing ketone body production.

The nurse caring for Mr. Nasser is responsible for monitoring his lipid panel, educating him on choosing unsaturated fats over saturated fats, and recognizing early signs of ketosis or ketoacidosis. The nurse also plays a key role in assessing for nausea, abdominal discomfort, or fruity breath odor that could indicate metabolic imbalance.

Understanding lipid metabolism allows the nurse to support the patient's nutritional goals, interpret lab results accurately, and collaborate with dietitians to tailor interventions. In this case, effective nursing care can help prevent complications and improve long-term outcomes.

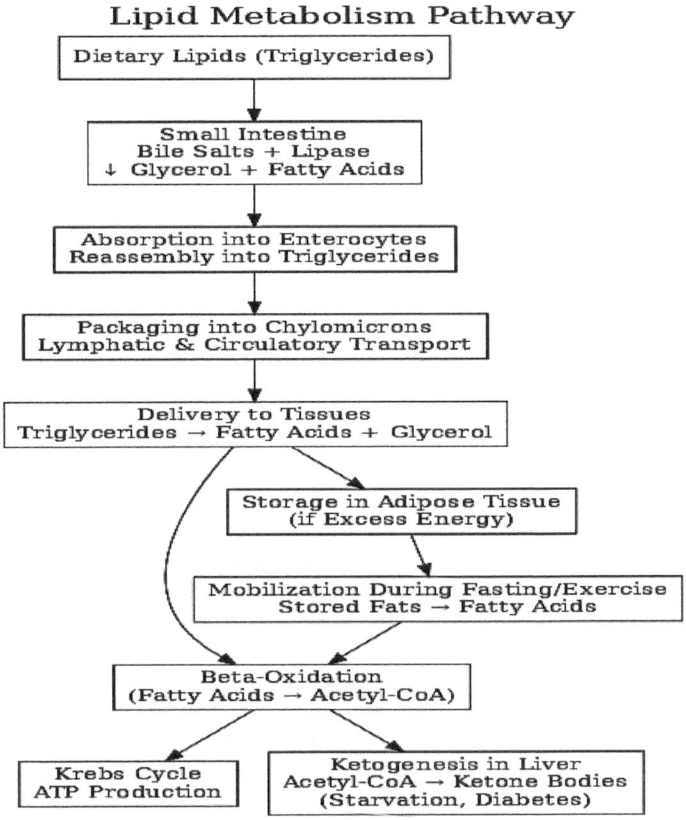

Figure 5.11 lipid Metabolic Pathway

Integration of Metabolism in the Human Body

Metabolic integration is the seamless coordination between the major nutrient pathways—carbohydrates, lipids, and proteins—to ensure the body's continuous supply of energy and building blocks across varying physiological states. Although each macronutrient is metabolized through distinct pathways, their end products converge at key metabolic hubs such as pyruvate, acetyl-CoA, and intermediates of the citric acid (Krebs) cycle. This convergence allows the body to flexibly adapt to changes in dietary intake, physical activity, stress, or illness while maintaining internal stability.

During the fed state, when nutrients are readily available, the anabolic hormone insulin predominates. It stimulates glucose uptake into cells, promotes glycogen synthesis in the liver and skeletal muscle, and activates lipogenesis in adipose tissue. Insulin also enhances protein synthesis by increasing amino acid uptake and reducing protein breakdown. Under these conditions, glucose serves as the primary energy source, and excess glucose is stored as glycogen or converted into fatty acids for long-term storage. This energy abundance enables the body to replenish energy reserves and support growth and tissue repair.

In contrast, during fasting or prolonged physical exertion, catabolic hormones such as glucagon and cortisol become dominant. These hormones signal the body to mobilize its stored energy reserves through glycogenolysis, lipolysis, and gluconeogenesis. Fatty acids released from adipose tissue are oxidized to produce ATP, while glucose is conserved for use by glucose-dependent tissues such as the brain and red blood cells. As glycogen stores become depleted, the liver converts fatty acids into ketone bodies—an alternative fuel that can cross the blood-brain barrier and partially substitute for glucose in the central nervous system.

Proteins, although not a primary energy source, may be broken down during prolonged starvation, illness, or metabolic stress. Amino acids released from muscle proteins can be used for gluconeogenesis or directly enter the Krebs cycle for energy production. However, protein catabolism is carefully regulated to preserve muscle function and structural integrity as long as possible. The nitrogen-containing waste products from amino acid deamination are converted to urea in the liver and excreted by the kidneys.

The liver plays a central role in integrating these metabolic pathways. It acts as a metabolic switchboard, processing dietary nutrients, managing energy reserves, and maintaining blood glucose levels through a balance of anabolic and catabolic reactions. By regulating substrate availability and hormone sensitivity, the liver ensures that energy demands are met under all physiological conditions.

This intricate interplay of metabolic processes is controlled by hormonal signals—primarily insulin, glucagon, epinephrine, and cortisol—as well as by the body's energy needs and nutrient availability. The ability to shift smoothly between energy storage and mobilization underpins the concept of homeostasis, enabling the body to function efficiently during feeding, fasting, stress, and recovery. For healthcare professionals, especially nurses, understanding these interconnections is essential for recognizing metabolic disturbances, managing nutritional therapy, and supporting patients in critical care settings.

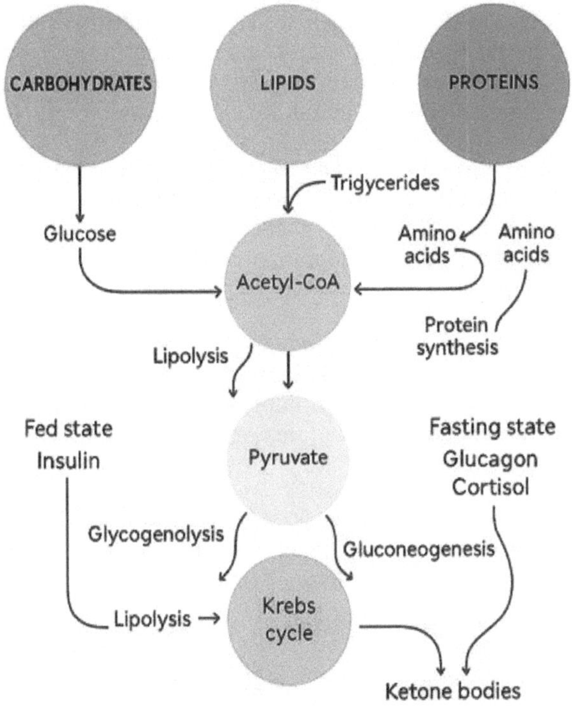

Figure 5.12 Integration of Metabolism Pathway

Chemistry Link to Health

The Biochemical Impact of Starvation in Times of Conflict

Starvation is a severe consequence of war, displacement, and economic collapse, especially in conflict-affected regions like Gaza, Sudan, and Yemen. It causes profound biochemical changes critical for healthcare providers to understand.

When food is scarce, glycogen stores are depleted within 24 hours, followed by fat breakdown into ketone bodies for energy. Once fat reserves are exhausted, muscle proteins are catabolized, causing muscle wasting, weakened immunity, and impaired organ function.

Clinically, serum albumin decreases, leading to edema; electrolyte imbalances affect cardiac and neurological function. Basal metabolic rate slows, and hormonal disruptions affect insulin, thyroid hormones, and cortisol. Vitamin and mineral deficiencies impair healing and immunity.

Recovery after refeeding is gradual. A 2024 study noted persistent intestinal inflammation and immune suppression in children; teduglutide showed promise in improving outcomes. Historical data confirm prolonged calorie restriction causes physical and psychological harm.

In conflict zones, starvation is a biochemical crisis. Nurses are crucial in monitoring, nutritional intervention, and compassionate care to manage and reverse these effects.

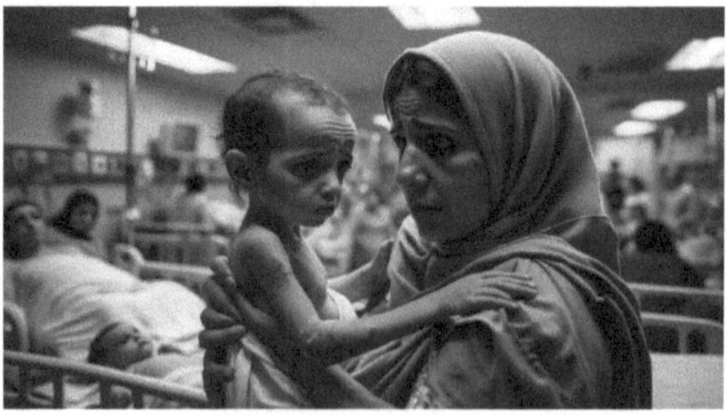

Chapter Summary

This chapter introduced the fundamental biomolecules essential for life—**carbohydrates, proteins, lipids, and nucleic acids**—highlighting their chemical structure, classification, biological roles, and clinical relevance.

Carbohydrates serve as the body's primary energy source. They include Monosaccharaides like glucose, disaccharides like lactose and sucrose, and polysaccharides like glycogen and cellulose. Nurses must understand carbohydrate metabolism in the context of diabetes, malnutrition, and digestion disorders.

Proteins are composed of amino acids and play structural, enzymatic, hormonal, and immune functions. Protein structure, from primary to quaternary levels, determines function. Clinical conditions such as enzyme deficiencies, autoimmune diseases, and protein-energy malnutrition are closely tied to protein metabolism.

Lipids include fatty acids, triglycerides, phospholipids, and steroids. They are essential for long-term energy storage, membrane structure, and hormone production. Disorders related to lipids—like hyperlipidemia and cardiovascular disease—are major concerns in nursing care and health education.

Nucleic acids (DNA and RNA) store and transmit genetic information and control protein synthesis. Understanding nucleic acids is foundational for interpreting genetic diseases, molecular diagnostics, and the principles of heredity.

The chapter concluded with an **overview of metabolism**, including the difference between catabolism and anabolism, and the central role of **ATP** in energy transfer. A simplified explanation of **glycolysis**, **Krebs cycle**, and **oxidative phosphorylation** prepared the groundwork for deeper metabolic studies.

By mastering the basic chemistry and biological functions of these macromolecules, nursing students are better equipped to understand disease mechanisms, interpret lab results, and provide informed, science-based patient care.

EXERCISES

Multiple Choice Questions (MCQs)

1. Which of the following is a disaccharide?
   A. Glucose
   B. Fructose
   C. Sucrose
   D. Cellulose
2. What type of bond links amino acids together in a protein?
   A. Hydrogen bond
   B. Peptide bond
   C. Glycosidic bond
   D. Ester bond
3. Which lipid is the precursor to steroid hormones?
   A. Triglyceride
   B. Phospholipid
   C. Cholesterol
   D. Fatty acid
4. What is the primary role of RNA in the cell?
   A. Storing genetic information
   B. Producing ATP
   C. Acting as a structural protein
   D. Assisting in protein synthesis
5. In which part of the cell does glycolysis occur?
   A. Nucleus
   B. Mitochondrial matrix
   C. Cytoplasm
   D. Endoplasmic reticulum

True or False

1. ------All amino acids are essential and must be obtained from the diet.
2. -----DNA contains uracil instead of thymine.
3. -----Lipids provide more energy per gram than carbohydrates.
4. -----ATP is produced only in the presence of oxygen.
5. -----Anabolism is a process that breaks down complex molecules into simpler ones.

Short Answer Questions

1. List the four major types of biomolecules and one function for each.
2. What are essential fatty acids, and why are they important?
3. Explain the role of ATP in metabolism.
4. Differentiate between catabolism and anabolism.
5. What is the function of enzymes in the human body?

Matching

Match each biomolecule in Column A with its corresponding function in Column

Column A (Biomolecule)	**Column B (Function/Feature)**
A. Glucose	1. Main structural component of cell membranes
B. DNA	2. Genetic material storing hereditary info
C. Phospholipid	3. Primary fuel for cellular respiration
D. Hemoglobin	4. Oxygen-carrying protein in red blood cells
E. Enzyme	5. Biological catalyst

Fill in the Blanks

1. The building blocks of proteins are called ______________.
2. The process of breaking down glucose into pyruvate is called ______________.
3. In DNA, adenine always pairs with ______________, while cytosine pairs with ______________.
4. The lipid responsible for hormone production and vitamin D synthesis is ______________.
5. A chain of monosaccharides forms a ______________.

Case studies

Case Study 1

Clinical Scenario:
Fatimah, a 58-year-old woman with a history of type 2 diabetes, presents to the clinic with complaints of fatigue, frequent urination, and blurred vision. Her laboratory results show a blood glucose level of 320 mg/dL and an HbA1c of 9.5%. Urinalysis reveals the presence of both glucose and ketones.

Questions for Discussion:

- What metabolic pathway is primarily impaired in this patient?
- What does the presence of ketones in the urine indicate about the patient's metabolic state?
- How might the metabolism of proteins be affected in a patient with poorly controlled diabetes?
- What nursing assessments and interventions are relevant in this case?

Case Study 2

Clinical Scenario:
Ahmad, a 6-year-old refugee child, is brought to a health clinic showing signs of lethargy, abdominal swelling, and dry skin. His diet consists mainly of white rice and sugary beverages. Lab tests reveal low serum albumin and evidence of muscle wasting.

Questions for Discussion:

- What type of nutritional deficiency is most likely affecting this child?
- What is the physiological role of albumin in the body?
- Why is albumin deficiency dangerous from a clinical perspective?
- What steps should nurses take to support nutritional recovery in such a case?

Case Study 3

Clinical Scenario:
Samer, a 45-year-old man with obesity, a history of smoking, and a family history of heart disease, comes in for a routine checkup. His lipid profile shows the following values:

- Total cholesterol: 260 mg/dL
- LDL: 180 mg/dL
- HDL: 35 mg/dL
- Triglycerides: 300 mg/dL

Questions for Discussion:

- Which component of the lipid profile is most concerning, and why?
- How does elevated LDL contribute to the pathophysiology of cardiovascular disease?
- What lifestyle and dietary recommendations should be considered for this patient?
- What nursing considerations are important when managing a patient at high cardiovascular risk?

Chapter 6

Bioactive Substances in the Body

Chapter 6: Bioactive Substances in the Body

Learning Objectives

By the end of this chapter, you should be able to:

1. Define bioactive substances and their role in human physiology.
2. Classify vitamins by solubility and describe key functions and sources.
3. Identify major hormone functions and related clinical disorders.
4. Explain the roles of key electrolytes in fluid, nerve, and muscle regulation.
5. Recognize signs of deficiencies or excesses in vitamins, hormones, and electrolytes.

The human body functions as a complex biochemical system regulated by vital bioactive substances—including vitamins, hormones, enzymes, and electrolytes—which collectively sustain homeostasis, metabolism, and cellular communication. While enzymes were discussed in detail in the previous chapter due to their foundational role in catalyzing metabolic reactions, this chapter will focus on the remaining categories: **vitamins, hormones, and electrolytes**. For nursing professionals, understanding these compounds is essential for interpreting lab values, managing nutritional and endocrine health, recognizing clinical symptoms of imbalance, and supporting therapeutic decisions. Hormones act as chemical messengers; electrolytes maintain nerve signaling and fluid balance; and vitamins serve as coenzymes and antioxidants critical to energy production, immunity, and tissue repair. Through structured explanations and real-life clinical examples, students will acquire the biochemical knowledge necessary for safe and holistic nursing care.

Vitamins

What Are Vitamins?

Vitamins are organic micronutrients required in small amounts for normal physiological processes, including growth, development, and metabolism. Unlike macronutrients, vitamins do not supply energy directly, but they function primarily as **coenzymes** or **cofactors** in essential biochemical pathways. Because the human body cannot synthesize most vitamins—or can produce them only in insufficient quantities—they must be obtained through the diet to maintain health.

Vitamins classification

Vitamins are classified into two major groups based on their solubility:

- **Water-soluble vitamins**: including the B-complex group and vitamin C. These dissolve in water, are not significantly stored in the body, and excess amounts are excreted in urine. They require regular dietary intake.
- **Fat-soluble vitamins**: including vitamins A, D, E, and K. These are absorbed along with dietary fats and stored in the liver and adipose tissue. While they do not need daily intake, excessive amounts can accumulate and cause toxicity.

This classification has direct **clinical relevance**, especially in evaluating nutritional deficiencies, prescribing supplementation, or managing malabsorption in nursing care.

Water-Soluble Vitamins

Water-soluble vitamins include the B-complex group and vitamin C. These vitamins dissolve easily in water, are not stored in significant amounts in the body, and excess is excreted through urine. As a result, they must be consumed regularly through the diet to maintain adequate physiological function. Despite their shared solubility, each vitamin has distinct roles in metabolism, nervous system function, red blood cell production, and immune support. The following overview outlines the individual importance of each water-soluble vitamin.

Vitamin B1 (Thiamine)

Thiamine functions as a coenzyme in carbohydrate metabolism, particularly in decarboxylation reactions within the Krebs cycle and the pentose phosphate pathway. It supports nerve conduction and energy generation. Deficiency impairs glucose utilization in the brain and nervous system, leading to beriberi and Wernicke-Korsakoff syndrome—especially prevalent in individuals with chronic alcohol abuse or malnutrition.

Vitamin B2 (Riboflavin)

Riboflavin is a key component of the coenzymes FAD (flavin adenine dinucleotide) and FMN (flavin mononucleotide), which are involved in electron transport and redox reactions. It contributes to cellular respiration, mucosal health, and antioxidant protection. Deficiency presents as ariboflavinosis, marked by sore throat, cheilitis, glossitis, and seborrheic dermatitis.

Vitamin B3 (Niacin)

Niacin exists as nicotinic acid and nicotinamide and forms the coenzymes NAD^+ and $NADP^+$, which are essential for oxidation-reduction reactions in cellular metabolism. The body can synthesize niacin from tryptophan, an essential amino acid. Severe deficiency leads to pellagra, a condition defined by the "three Ds": dermatitis, diarrhea, and dementia. High therapeutic doses of niacin may be used to treat hyperlipidemia but can cause flushing, liver toxicity, and gastrointestinal distress.

Vitamin B6 (Pyridoxine)

Pyridoxine and its active form, pyridoxal phosphate, serve as coenzymes in amino acid metabolism, neurotransmitter synthesis (such as serotonin, dopamine, and GABA), glycogenolysis, and immune function. Deficiency can result in irritability, peripheral neuropathy, depression, and microcytic anemia. Supplementation beyond recommended levels may lead to irreversible nerve damage.

Folic Acid (Vitamin B9)

Folic acid is essential for DNA and RNA synthesis, cell division, and amino acid metabolism. It is particularly important during periods of rapid cell proliferation, such as pregnancy. Deficiency causes megaloblastic anemia and increases the risk of neural tube defects in the fetus. Although folic acid is non-toxic at typical dietary levels, high doses can mask the hematological symptoms of vitamin B12 deficiency, potentially delaying its diagnosis.

Vitamin B12 (Cobalamin)

Vitamin B12 plays a critical role in red blood cell formation, DNA synthesis, and the maintenance of myelin in the nervous system. It requires intrinsic factor, secreted by the gastric mucosa, for absorption in the ileum. It is found exclusively in animal-based foods, making vegetarians and individuals with malabsorption syndromes more susceptible to deficiency. Deficiency leads to megaloblastic anemia, fatigue, and potentially irreversible neurological damage, including paresthesia and memory loss.

Vitamin C (Ascorbic Acid)

Vitamin C is a water-soluble antioxidant that protects cells from oxidative stress and supports the immune system. It is also required for the hydroxylation of proline and lysine during collagen synthesis, which is essential for wound healing, skin integrity, and capillary strength. Additionally, it enhances the absorption of non-heme iron in the gastrointestinal tract. Deficiency results in scurvy, characterized by bleeding gums, petechiae, joint pain, and delayed wound healing. Although vitamin C is generally well-tolerated, excessive intake may lead to gastrointestinal discomfort and increase the risk of kidney stones in predisposed individuals.

Clinical Relevance in Nursing Practice

Water-soluble vitamin deficiencies are commonly encountered in patients with malnutrition, chronic illness, gastrointestinal disorders, or alcoholism. Nurses play a crucial role in recognizing deficiency symptoms, monitoring dietary intake, interpreting laboratory findings, and administering appropriate supplementation. Moreover, understanding the biochemical roles of these vitamins equips nurses to educate patients on nutritional requirements and the safe use of multivitamin therapy.

Fat-Soluble Vitamins

Fat-soluble vitamins—**A, D, E, and K**—are absorbed in the intestines along with dietary fats and require bile acids for proper digestion and absorption. Unlike water-soluble vitamins, they are stored in the liver and adipose tissue, which reduces the need for daily intake but increases the risk of toxicity when consumed in excess. These vitamins play diverse yet essential roles in vision, bone health, antioxidant defense, and blood coagulation.

Vitamin A (Retinol and Beta-Carotene)

Vitamin A plays a central role in maintaining healthy vision, particularly under low-light conditions, by participating in the formation of rhodopsin, a light-sensitive pigment in the retina. It also supports immune system function, promotes the health of epithelial tissues such as the skin and mucous membranes, and contributes to normal growth and reproduction. Dietary sources of preformed vitamin A (retinol) include liver, dairy products, and butter, while provitamin A carotenoids—such as beta-carotene—are found in orange and dark green vegetables like carrots, sweet potatoes, and spinach. A deficiency in vitamin A may lead to night blindness, dry eyes (xerophthalmia), keratinization of the skin, and increased susceptibility to infections. However, excessive intake of vitamin A, especially from high-dose supplements, can cause toxic effects such as liver damage, birth defects during pregnancy (teratogenicity), and increased intracranial pressure (pseudotumor cerebri). For nursing professionals, it is essential to monitor vitamin A intake, particularly in pregnant women and individuals with liver conditions, and to educate patients on achieving an appropriate balance between dietary sources and supplementation.

Vitamin D (Calciferol)

Vitamin D is a fat-soluble prohormone that is crucial for regulating calcium and phosphate metabolism, thereby ensuring proper bone mineralization and neuromuscular function. It enhances the intestinal absorption of calcium and phosphate and modulates immune responses. Vitamin D is synthesized in the skin through exposure to ultraviolet B (UVB) rays from sunlight, and it is also obtained from dietary sources such as fatty fish, fortified milk, and egg yolks. A deficiency in vitamin D can lead to rickets in children, characterized by bone deformities and delayed growth, or osteomalacia in adults, presenting with bone pain, muscle weakness, and increased fracture risk. On the other hand, excessive supplementation of vitamin D can result in hypercalcemia, kidney stone formation, and calcification of soft tissues, including the heart and blood vessels. For nurses, it is important to assess patients' risk factors for deficiency—such as limited sun exposure, dark skin, obesity, or malabsorption syndromes—and to educate patients on the safe use of supplements and the importance of balanced sun exposure and nutrition.

Vitamin E (Tocopherol)

Vitamin E acts as a powerful lipid-soluble antioxidant that protects cell membranes from oxidative stress by neutralizing free radicals. It contributes to immune defense, cellular signaling, and the preservation of neurological function. Major dietary sources of vitamin E include vegetable oils such as sunflower and safflower oil, as well as nuts, seeds, and green leafy vegetables. While vitamin E deficiency is rare, it may occur in individuals with fat malabsorption syndromes (e.g., cystic fibrosis or cholestasis), leading to symptoms such as neuromuscular problems, peripheral neuropathy, impaired coordination (ataxia), and hemolytic anemia, particularly in premature infants. Although vitamin E is considered safe at moderate dietary levels, high-dose supplementation may increase the risk of bleeding due to interference with vitamin K–dependent clotting factors. Nurses should be vigilant when caring for patients who take anticoagulants, such as warfarin or aspirin, and should provide counseling on the appropriate use of vitamin E supplements, emphasizing food sources over excessive supplementation.

Vitamin K (Phylloquinone and Menaquinone)

Vitamin K is essential for the post-translational modification of specific proteins involved in blood coagulation and bone metabolism. It facilitates the carboxylation of glutamic acid residues on clotting factors II, VII, IX, and X, allowing them to bind calcium and function effectively. Vitamin K1 (phylloquinone) is found in green leafy vegetables such as kale, spinach, and broccoli, while vitamin K2 (menaquinone) is synthesized by intestinal microbiota. Deficiency in vitamin K can lead to prolonged bleeding, easy bruising, and hemorrhagic disease of the newborn. It is particularly concerning in individuals taking long-term antibiotics, those with liver or gastrointestinal disorders, and in neonates who have insufficient stores at birth. Although toxicity from vitamin K is rare, especially from dietary sources, high intake can antagonize anticoagulant therapy, such as warfarin, reducing its effectiveness. Nurses should ensure newborns receive vitamin K injections to prevent bleeding complications and should guide patients on anticoagulant therapy to maintain a consistent intake of vitamin K–rich foods to avoid fluctuations in medication efficacy.

Clinical Relevance in Nursing Practice

For nurses, recognizing the clinical impact of fat-soluble vitamins is essential when assessing nutritional status, managing supplementation, and preventing complications. Because these vitamins are stored in body tissues, both deficiencies and toxicities can have delayed but serious effects. Nursing responsibilities include identifying at-risk groups (e.g., individuals with malabsorption, restricted diets, or on specific medications), monitoring for signs of imbalance, ensuring appropriate prophylaxis (such as vitamin K in newborns), and providing clear patient education on safe intake, especially when supplements or anticoagulants are involved. Integrating this knowledge enhances patient safety and supports evidence-based care.

The following tables include a summary of essential and non-essential vitamins required by the body, outlining their vital functions, dietary sources, deficiency symptoms, and potential toxicity risks. This classification aims to facilitate understanding of the importance of each vitamin and its role in maintaining the body's health.

Table 6.1 Water soluble Vitamins Summary

Vitamin	Sources	Key Functions	Deficiency Effects	Toxicity/ Excess
B1	Whole grains, legumes, pork	Carbohydrate metabolism, nerve function	Beriberi, Wernicke-Korsakoff syndrome	Rare
B2	Milk, eggs, leafy greens, meat	Redox reactions, energy metabolism	Cheilitis, glossitis, sore throat	Rare
B3	Meat, fish, whole grains, legumes	Energy transfer (NAD/NADP)	Pellagra: dermatitis, diarrhea, dementia	Flushing, liver damage
B6	Poultry, bananas, potatoes, cereals	Amino acid metabolism, neurotransmitters	Neuropathy, anemia, depression	Nerve damage in high doses
B9	Leafy greens, citrus fruits, legumes	DNA synthesis, fetal development	Megaloblastic anemia, neural tube defects	May mask B12 deficiency
B12	Meat, dairy, eggs	RBC production, nerve function	Pernicious anemia, neuropathy	None known
C	Citrus fruits, peppers, strawberries	Antioxidant, collagen synthesis, iron absorption	Scurvy: bleeding gums, poor healing	Diarrhea, kidney stones

Table 6.2 Fat soluble Vitamins Summary

Vitamin	**Sources**	**Key Functions**	**Deficiency Effects**	**Toxicity/ Excess**
A	Liver, dairy, carrots, sweet potatoes	Vision, immunity, epithelial integrity	Night blindness, infection risk	Liver toxicity, birth defects
D	Sunlight, fatty fish, fortified milk	Calcium/phosphate balance, bone health	Rickets, osteomalacia	Hypercalcemia, kidney stones
E	Vegetable oils, nuts, seeds, greens	Antioxidant, protects cell membranes	Neurological issues, muscle weakness	Bleeding risk (with anticoagulants)
K	Leafy greens, gut bacteria	Blood clotting, bone health	Bleeding tendency, especially in newborns	Rare, interacts with anticoagulants

Figure 6.1 Natural Sources of Vitamins

Case Study:

Mrs. Amal, a 30-year-old pregnant woman in her first trimester, visits the antenatal clinic complaining of persistent fatigue and occasional dizziness. She follows a low-vegetable, high-starch diet and admits to skipping prenatal supplements due to nausea. Her blood tests reveal macrocytic anemia, and fetal ultrasound shows early neural tube development concerns. The attending nurse suspects a nutritional deficiency and reviews her dietary intake and lab results in detail.

Nursing Analysis and Intervention:

Based on the symptoms and laboratory findings, Mrs. Amal is likely suffering from **folic acid (Vitamin B9) deficiency**, which is critical during early pregnancy for proper **DNA synthesis and fetal neural tube development**. This case reflects the "Deficiency Effects" outlined in Table 6.1, including **megaloblastic anemia and neural tube defects**. The nurse's intervention includes initiating folic acid supplementation immediately, providing dietary education focused on leafy greens, citrus fruits, and legumes, and collaborating with the physician to adjust prenatal care. Early correction can significantly improve maternal energy levels and reduce fetal complication risks.

Hormones

Hormones are chemical messengers produced by endocrine glands and released into the bloodstream, where they regulate essential physiological processes including growth, metabolism, reproduction, fluid and electrolyte balance, and stress responses. Although they are secreted in minute quantities, hormones exert significant effects on target tissues by binding to specific receptors and initiating complex biochemical cascades that maintain internal homeostasis.

Unlike nutrients, hormones do not provide structural components or energy, but they serve as regulators that coordinate body systems. Their imbalances—whether due to overproduction, underproduction, or receptor malfunction—can result in serious medical conditions. Disorders such as diabetes mellitus, hypothyroidism, Addison's disease, and Cushing's syndrome arise from hormonal dysregulation and require vigilant assessment, ongoing treatment, and patient education.

For nursing professionals, a solid understanding of hormonal function and clinical manifestations of imbalance is crucial for effective patient care. Nurses play a key role in interpreting laboratory results, administering hormone-related therapies, and educating patients about chronic endocrine conditions. The following table summarizes the sources, physiological roles, deficiency and excess effects, and clinical implications of major human hormones.

Insulin

Insulin is a peptide hormone produced by the β-cells of the pancreas within the islets of Langerhans. It is secreted in response to elevated blood glucose levels following carbohydrate intake. Insulin facilitates the cellular uptake of glucose, particularly in muscle, adipose, and liver tissues, where it promotes its conversion into glycogen and fat for storage. This action results in a decrease in circulating blood glucose, helping maintain glucose homeostasis. A deficiency in insulin production or a reduced cellular response to insulin—known as insulin resistance—leads to hyperglycemia and forms the basis of diabetes mellitus. In **Type 1 diabetes**, the body's immune system attacks and destroys pancreatic β-cells, leading to absolute insulin deficiency. In **Type 2 diabetes**, insulin is still produced but tissues become resistant to its effects.

Conversely, excessive insulin levels, often due to therapeutic administration, can cause **hypoglycemia**, manifesting as sweating, dizziness, confusion, palpitations, and, in severe cases, loss of consciousness. For nurses, understanding insulin's mechanism and recognizing signs of hypo- or hyperglycemia is vital for safe medication administration and patient education.

Glucagon

Glucagon is a peptide hormone secreted by the α-cells of the pancreas in response to low blood glucose levels, particularly during fasting or hypoglycemia. It acts in physiological opposition to insulin by increasing blood glucose concentrations. Glucagon achieves this by stimulating **glycogenolysis** (the breakdown of glycogen into glucose) and **gluconeogenesis** (the synthesis of glucose from non-carbohydrate precursors) in the liver. These actions help maintain blood glucose levels during periods of energy demand or carbohydrate deprivation. While glucagon deficiency is rare, it can impair the body's ability to recover from hypoglycemic episodes.

Conversely, excessive glucagon production—such as in glucagon-secreting pancreatic tumors (glucagonomas)—may result in **hyperglycemia**, **unintended weight loss**, and other metabolic disturbances. For nurses, understanding the balance between insulin and glucagon is vital for managing glycemic control in patients, especially those with diabetes or endocrine tumors.

Thyroxine (T4) and Triiodothyronine (T3)

Thyroxine (T4) and triiodothyronine (T3) are the principal thyroid hormones synthesized and secreted by the thyroid gland under the regulatory influence of thyroid-stimulating hormone (TSH), which is released by the anterior pituitary gland. These hormones are critical regulators of the body's **basal metabolic rate**, influencing oxygen consumption, energy expenditure, heart rate, thermoregulation, and the development of the central nervous system. Iodine is an essential element required for the synthesis of both T4 and T3. A deficiency in these hormones leads to **hypothyroidism**, which may present clinically with symptoms such as fatigue, weight gain, cold intolerance, constipation, dry skin, and bradycardia. In infants and young children, severe hypothyroidism can result in **cretinism**, a condition marked by growth retardation and intellectual disability.

Conversely, excessive production of thyroid hormones causes **hyperthyroidism**, characterized by weight loss despite normal or increased appetite, heat intolerance, anxiety, tremors, tachycardia, and insomnia. For nurses, recognizing these opposing clinical pictures is essential for supporting accurate diagnosis, medication management (e.g., levothyroxine or antithyroid drugs), and patient counseling.

Cortisol

Cortisol is a steroid hormone belonging to the glucocorticoid class, secreted by the adrenal cortex in response to adrenocorticotropic hormone (ACTH) released from the anterior pituitary. It plays a vital role in the body's **stress response**, **regulation of glucose metabolism**, **maintenance of blood pressure**, and **suppression of inflammation and immune responses**. Cortisol promotes gluconeogenesis, mobilizes energy reserves, and enhances cardiovascular tone during stress. Its secretion follows a diurnal rhythm, peaking in the early morning and increasing further during physical or emotional stress. Chronic excess of cortisol results in **Cushing's syndrome**, a condition characterized by central obesity, a rounded "moon face," thin limbs due to muscle wasting, hypertension, and impaired wound healing.

On the other hand, cortisol deficiency leads to **Addison's disease**, which presents with fatigue, low blood pressure, weight loss, salt craving, and skin hyperpigmentation due to compensatory elevation of ACTH. For nursing professionals, monitoring cortisol-related disorders is essential for ensuring safe administration of corticosteroids, interpreting endocrine tests, and educating patients on stress management and medication adherence.

Adrenaline (epinephrine)

Adrenaline, also known as epinephrine, is a catecholamine hormone synthesized and secreted by the adrenal medulla in response to sympathetic nervous system activation, particularly during acute stress or the "fight or flight" response. It prepares the body for immediate action by **increasing heart rate and cardiac output**, **dilating the bronchioles** to enhance oxygen intake, **raising blood glucose levels** through glycogenolysis, and **redirecting blood flow** away from non-essential organs toward skeletal muscles. These rapid physiological changes improve physical performance and survivability in emergency situations.

Overproduction of adrenaline may occur in cases of **pheochromocytoma**, a rare tumor of the adrenal medulla, leading to **hypertension**, **palpitations**, **headaches**, and **anxiety-like symptoms**. True adrenaline deficiency is rare due to redundancy in the body's stress response systems, but impaired secretion may blunt the body's ability to respond effectively to acute stressors such as severe hypoglycemia or shock. For nurses, recognizing symptoms of excess catecholamines and managing emergency medications such as epinephrine injections is critical in acute care settings..

Aldosterone

Aldosterone is a mineralocorticoid hormone secreted by the adrenal cortex as part of the renin–angiotensin–aldosterone system (RAAS). It plays a critical role in regulating **sodium and potassium balance**, as well as **maintaining blood volume and blood pressure**. Aldosterone acts primarily on the distal tubules and collecting ducts of the kidneys, where it promotes **sodium reabsorption** and **potassium excretion**. This sodium retention leads to water retention, thereby supporting circulatory volume and hemodynamic stability. **Aldosterone deficiency**, as seen in **Addison's disease**, results in **hyponatremia**, **hyperkalemia**, **hypotension**, dehydration, and fatigue.

On the other hand, **excess aldosterone secretion**, such as in **Conn's syndrome** (primary hyperaldosteronism), causes **hypertension**, **hypokalemia**, and **metabolic alkalosis**. For nurses, monitoring electrolyte levels, blood pressure, and signs of fluid imbalance is essential in patients with suspected adrenal dysfunction, and patient education on salt balance and medication adherence plays a key role in long-term management.

Parathyroid Hormone (PTH)

Parathyroid hormone (PTH) is a peptide hormone secreted by the parathyroid glands in response to low serum calcium levels. It plays a critical role in maintaining **calcium and phosphate homeostasis** in the body. PTH increases blood calcium by stimulating **bone resorption**, enhancing **renal calcium reabsorption**, and activating **vitamin D (calcitriol)**, which in turn increases intestinal calcium absorption. At the same time, PTH decreases phosphate reabsorption in the kidneys, promoting phosphate excretion. **Deficiency of PTH**, known as **hypoparathyroidism**, leads to **hypocalcemia**, which may present with **muscle cramps**, **numbness**, **tingling**, and **tetany**—a condition of involuntary muscle contractions.

In contrast, **excessive secretion of PTH**, or **hyperparathyroidism**, results in **hypercalcemia**, which can cause **bone demineralization**, **osteoporosis**, **kidney stones**, **fatigue**, and **abdominal pain**. For nurses, recognizing signs of calcium imbalance and understanding the relationship between PTH, vitamin D, and kidney function is crucial in managing endocrine and renal patients.

Antidiuretic Hormone (ADH or vasopressin)

Antidiuretic hormone (ADH), also known as vasopressin, is synthesized by the hypothalamus and stored and released by the posterior pituitary gland. It plays a crucial role in regulating the body's **fluid balance and blood pressure** by acting on the **collecting ducts of the kidneys** to promote **water reabsorption**. By conserving water, ADH helps concentrate the urine and maintain circulatory volume. A **deficiency of ADH**, as seen in **diabetes insipidus**, leads to the excretion of large volumes of dilute urine (polyuria), excessive thirst (polydipsia), and risk of **dehydration** and **electrolyte imbalance**.

In contrast, **excessive ADH secretion**, such as in **Syndrome of Inappropriate ADH Secretion (SIADH)**, causes **water retention**, **dilutional hyponatremia**, and symptoms such as headache, confusion, nausea, and in severe cases, seizures or coma. For nurses, early recognition of ADH-related imbalances is vital in managing fluid therapy, monitoring electrolytes, and preventing complications related to overhydration or dehydration.

Oxytocin

Oxytocin is a peptide hormone synthesized in the hypothalamus and released by the posterior pituitary gland. It plays a key role in reproductive physiology, particularly by stimulating uterine contractions during labor and promoting milk ejection (let-down reflex) during breastfeeding. In addition to its physiological effects, oxytocin is also implicated in maternal-infant bonding, emotional regulation, and social behavior, earning it the nickname "the bonding hormone." Clinically, synthetic oxytocin (commonly known by the trade name *Pitocin*) is used to induce or augment labor, as well as to reduce postpartum hemorrhage by enhancing uterine tone.

While oxytocin deficiency may result in poor milk ejection and challenges with breastfeeding, excessive levels are generally not harmful in most individuals due to the hormone's rapid clearance from circulation. For nurses, understanding the role of oxytocin is important in labor and delivery units, maternal education, and newborn care.

Growth Hormone (GH)

Growth hormone (GH), also known as somatotropin, is a peptide hormone secreted by the anterior pituitary gland under the regulation of **growth hormone-releasing hormone (GHRH)** and **somatostatin** from the hypothalamus. GH plays a central role in **stimulating linear growth in children**, as well as promoting **protein synthesis**, **lipid metabolism**, and **cellular regeneration** in individuals of all ages. In children and adolescents, **GH deficiency** results in **dwarfism**, characterized by short stature and delayed development, whereas **excess GH** leads to **gigantism**, marked by abnormal height due to excessive bone growth before epiphyseal plate closure.

In adults, continued GH overproduction causes **acromegaly**, a condition distinguished by **enlargement of the hands, feet, jaw, and facial bones**, along with metabolic disturbances such as insulin resistance and cardiovascular complications. For nurses, understanding GH function and its clinical implications supports early detection of growth disorders, monitoring of GH therapy, and patient education on associated risks and treatments.

Prolactin

Prolactin is a peptide hormone produced by the anterior pituitary gland that stimulates breast development and milk production. Its levels naturally rise during pregnancy and breastfeeding. Excess prolactin, a condition known as hyperprolactinemia, can lead to infertility, menstrual irregularities, and spontaneous milk secretion (galactorrhea), often due to pituitary adenomas or side effects of certain medications. Deficiency is uncommon and typically results only in impaired lactation.

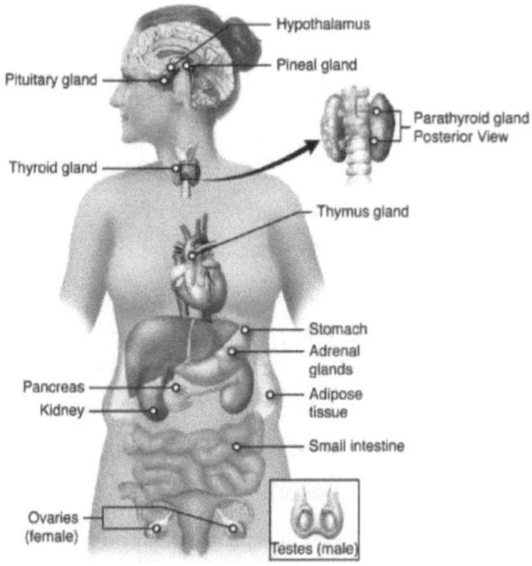

Figure 6.2 The Endocrine System Glands

Table 6.3 Pancreatic and Adrenal Hormones

Hormone	Source	Main Function	Deficiency	Excess
Insulin	Pancreas (beta cells)	Lowers blood glucose, promotes storage	Diabetes mellitus (Type 1 or 2)	Hypoglycemia (from overdose)
Glucagon	Pancreas (alpha cells)	Raises blood glucose during fasting	Impaired glucose response	Hyperglycemia (rare tumors)
Cortisol	Adrenal cortex	Stress response, glucose metabolism	Addison's disease	Cushing's syndrome
Adrenaline	Adrenal medulla	Fight-or-flight: ↑HR, ↑glucose	Poor stress response	Pheochromocytoma
Aldosterone	Adrenal cortex	Sodium retention, K^+ excretion	Addison's disease	Conn's syndrome (HTN, ↓K^+)

Table 6.4: Hormones from Other Glands

Hormone	Source	Main Function	Deficiency	Excess
Thyroxine (T4), T3	Thyroid gland	Regulate metabolism and growth	Hypothyroidism	Hyperthyroidism
Parathyroid Hormone	Parathyroid glands	Increases blood calcium	Hypocalcemia, tetany	Hypercalcemia, bone loss
ADH (Vasopressin)	Hypothalamus / Posterior pituitary	Water retention, concentrates urine	Diabetes insipidus	SIADH (fluid retention)
Oxytocin	Posterior pituitary	Uterine contractions, milk ejection	Poor lactation	Rarely problematic
Growth Hormone (GH)	Anterior pituitary	Stimulates growth, protein synthesis	Dwarfism in children	Gigantism / Acromegaly
Prolactin	Anterior pituitary	Milk production	Poor milk supply	Infertility, galactorrhea

Case Study:

Mr. Tarek, a 47-year-old male, presents to the emergency department with complaints of muscle weakness, frequent urination, extreme thirst, and unexplained weight loss. On examination, his blood pressure is elevated, and laboratory tests reveal hypokalemia (low potassium), hypernatremia (high sodium), and metabolic alkalosis. Further evaluation shows elevated plasma aldosterone and suppressed renin levels. The endocrinology team suspects an adrenal-related disorder affecting electrolyte balance.

Nursing Analysis and Intervention:

Mr. Tarek's clinical picture is consistent with primary hyperaldosteronism (Conn's syndrome), a condition characterized by excess aldosterone secretion from the adrenal cortex. According to Table 6.2, aldosterone promotes sodium retention and potassium excretion, which explains the patient's hypertension and hypokalemia. The nurse plays a vital role in monitoring vital signs, cardiac rhythm (due to electrolyte imbalance), and educating the patient about sodium-restricted diets and medication adherence. Coordination with the medical team for imaging studies and possible surgical referral (if an adrenal tumor is found) is essential for definitive management.

Electrolytes and Ions

Electrolytes are electrically charged particles, or ions, that dissolve in the body's fluids and are essential for maintaining numerous physiological processes. These include fluid and osmotic balance, nerve impulse transmission, muscle contraction, acid–base regulation, and enzymatic reactions. The most clinically significant electrolytes are sodium (Na^+), potassium (K^+), calcium (Ca^{2+}), magnesium (Mg^{2+}), chloride (Cl^-), phosphate (PO_4^{3-}), and bicarbonate (HCO_3^-). Their concentrations are precisely controlled by the renal system, hormonal signals (such as aldosterone and ADH), and buffer systems in the blood. Electrolyte imbalances can result from various causes, including dehydration, kidney dysfunction, endocrine disorders, medications, or acute illnesses. Clinically, these disturbances may manifest as muscle weakness, arrhythmias, confusion, seizures, or shock. For nurses, the ability to assess electrolyte values, interpret laboratory data, and promptly identify clinical signs of imbalance is essential for safe and effective patient care.

Key Electrolytes – Clinical Descriptions

Sodium (Na^+)

Sodium (Na^+) is primarily obtained from dietary sources such as **table salt (sodium chloride), processed and canned foods, cheese, snack products**, and **salted meats**. As the **main extracellular cation**, sodium is vital for maintaining **fluid distribution**, **blood pressure regulation**, and **nerve impulse transmission**. It plays a central role in creating **osmotic gradients** between intracellular and extracellular compartments and contributes to the **acid–base balance**.

Sodium homeostasis is tightly regulated by two major hormones: **aldosterone**, which increases sodium reabsorption in the renal tubules, and **antidiuretic hormone (ADH)**, which modulates water reabsorption and thus influences sodium concentration through dilution or concentration.

Hyponatremia (low sodium levels) can result from excessive fluid intake, heart or kidney failure, or conditions like **Syndrome of Inappropriate ADH Secretion (SIADH)**. Clinically, it may present with **nausea, headache, confusion, muscle cramps, seizures**, and **coma** in severe cases.

On the other hand, **hypernatremia** (elevated sodium levels), often due to **dehydration**, **prolonged diarrhea**, or **diabetes insipidus**, may cause **intense thirst, restlessness, irritability**, and **neurologic disturbances**.

In nursing practice, accurate **assessment of fluid status**, **interpretation of laboratory sodium values**, and **adjustment of IV fluids and electrolyte therapy** are crucial to preventing complications and ensuring optimal patient outcomes.

Potassium (K^+)

Potassium (K^+) is abundant in foods such as **bananas, potatoes, oranges, spinach, avocados, tomatoes, legumes**, and **dairy products**. It is the **primary intracellular cation**, playing a central role in **maintaining the resting membrane potential** necessary for **nerve conduction, muscle contraction**, and **cardiac rhythm regulation**.

Potassium levels in the body are tightly regulated by the **kidneys**, which respond to aldosterone and other hormonal signals to excrete or conserve potassium as needed. **Acid–base status, cellular shifts**, and **medications**—especially **diuretics, insulin**, and **ACE inhibitors**—also influence potassium balance.

Hypokalemia (low potassium) may result from vomiting, diarrhea, diuretic use, or alkalosis, and can cause **muscle weakness, fatigue, constipation, arrhythmias**, and characteristic **ECG changes** such as flattened T waves and the appearance of U waves. In contrast, **hyperkalemia** (high potassium) often arises in cases of renal failure or medication effects, and poses a serious risk for **life-threatening cardiac arrhythmias** and **asystole**.

For nurses, monitoring potassium levels, recognizing early clinical signs, adjusting medications, and administering potassium supplements or binders are essential components of safe electrolyte management.

Calcium (Ca^{2+})

Calcium (Ca^{2+}) is primarily obtained through dietary sources such as **milk and dairy products, leafy green vegetables** (e.g., kale, broccoli), **fortified cereals, tofu, almonds**, and **fish with edible bones** (like sardines). It is the most abundant mineral in the human body, with over 99% stored in **bones and teeth**, where it provides structural strength.

Biologically, calcium is essential for **muscle contraction, nerve signal transmission, blood clotting, hormone secretion**, and **enzyme activation**. In circulation, calcium exists in both **bound (to**

albumin) and **ionized (free)** forms—the latter being physiologically active.

Calcium homeostasis is finely regulated by **parathyroid hormone (PTH)**, which increases calcium levels by promoting bone resorption, enhancing intestinal absorption via **activation of vitamin D**, and reducing renal excretion. **Calcitonin**, secreted by the thyroid, lowers blood calcium by inhibiting bone breakdown.

Hypocalcemia (low calcium) may result from **vitamin D deficiency**, **hypoparathyroidism**, **renal disease**, or **acute pancreatitis**, and manifests clinically as **tetany**, **muscle cramps**, **perioral numbness**, **laryngospasm**, and **cardiac arrhythmias** (prolonged QT interval). In contrast, **hypercalcemia**, which can occur in **hyperparathyroidism** or **malignancy**, leads to **fatigue, constipation, kidney stones, polyuria**, **bone pain**, and **confusion**.

Nursing responsibilities include **monitoring serum calcium (total and ionized)**, recognizing neuromuscular and cardiac signs, administering calcium or vitamin D supplements, and evaluating potential underlying causes.

Magnesium (Mg^{2+})

Magnesium (Mg^{2+}) is abundantly found in **green leafy vegetables** (such as spinach), **nuts** (like almonds and cashews), **whole grains**, **legumes**, **seeds**, **avocados**, and **dark chocolate**. It is an essential intracellular cation that plays a critical role in **over 300 enzymatic reactions**, including those involved in **protein synthesis**, **ATP production**, **DNA and RNA stability**, and **neuromuscular transmission**.

Magnesium supports **normal muscle and nerve function**, helps regulate **cardiac rhythm**, and contributes to **bone structure** and **glucose metabolism**. It also serves as a natural **calcium channel blocker**, moderating smooth muscle tone and neuronal excitability.

Hypomagnesemia (low magnesium), often caused by **malnutrition**, **chronic alcoholism**, **diuretic therapy**, or **gastrointestinal losses**, presents with **tremors**, **tetany**, **muscle cramps**, **positive Chvostek's**

and Trousseau's signs, **seizures**, and **ventricular arrhythmias**, particularly in critically ill patients.

Hypermagnesemia (high magnesium), although less common, typically arises in the context of **renal insufficiency**, **excessive magnesium-containing antacids or laxatives**, and may result in **flushing, hypotension, bradycardia, diminished deep tendon reflexes**, **respiratory depression**, and **cardiac arrest** in severe cases.

From a nursing perspective, it is essential to monitor magnesium levels in patients receiving IV fluids, diuretics, or magnesium-based medications, and to promptly recognize symptoms of imbalance to prevent complications.

Chloride (Cl^-)

Chloride (Cl^-) is widely available in the diet, primarily through **table salt (sodium chloride)**, as well as in **processed foods**, **cheese**, **seaweed**, **tomatoes**, **celery**, and **olives**. As the **principal extracellular anion**, chloride plays a fundamental role in **maintaining osmotic pressure**, **electrical neutrality**, and **acid–base homeostasis**.

One of its key physiological roles is contributing to the formation of **hydrochloric acid (HCl)** in the stomach, which is essential for **protein digestion** and **innate immunity**. Chloride also participates in **chloride–bicarbonate exchange** across red blood cells, which helps regulate **blood pH**.

Hypochloremia (low chloride levels) commonly occurs alongside **metabolic alkalosis**, **prolonged vomiting**, **diuretic therapy**, or **excessive gastric suction**, and may present with **muscle weakness**, **tetany**, **shallow breathing**, and **hypoventilation**.

Hyperchloremia (high chloride levels), on the other hand, is often associated with **dehydration**, **renal dysfunction**, or **metabolic acidosis**, and may lead to **lethargy**, **deep rapid breathing (Kussmaul respirations)**, and **altered mental status**.

For nurses, careful monitoring of chloride levels—especially in patients receiving IV fluids, suffering from acid–base disorders, or undergoing diuretic therapy—is critical for maintaining homeostasis and preventing further electrolyte imbalance.

Phosphate (PO_4^{3-})

Phosphate (PO_4^{3-}) is abundant in **meat, fish, dairy products, eggs, legumes, nuts**, and **whole grains**. It is a vital component of **adenosine triphosphate (ATP)**, the energy currency of the cell, and plays key roles in **bone and teeth mineralization, buffering of acids and bases, cell membrane integrity (phospholipids)**, and **intracellular signaling pathways**.

Approximately 85% of the body's phosphate is stored in bones, while the remainder is involved in cellular metabolism and enzymatic reactions. Phosphate levels are regulated primarily by the **kidneys**, and influenced by **parathyroid hormone (PTH)**, **vitamin D**, and **fibroblast growth factor 23 (FGF23)**.

Hypophosphatemia (low phosphate) may result from **malnutrition, chronic alcoholism, diabetic ketoacidosis**, or **refeeding syndrome**. Clinically, it manifests as **muscle weakness, bone pain, rhabdomyolysis, impaired diaphragmatic contraction**, and in severe cases, **respiratory failure**.

Hyperphosphatemia (elevated phosphate), often seen in **chronic kidney disease**, is associated with **calcium-phosphate precipitation**, which may lead to **vascular calcification, secondary hyperparathyroidism**, and **skeletal abnormalities**.

Nurses should be vigilant in monitoring phosphate levels in patients at risk, managing dietary intake, and recognizing symptoms related to bone, muscle, and metabolic function.

Bicarbonate (HCO_3^-)

Bicarbonate (HCO_3^-) is not a major dietary electrolyte, but small amounts are found in **baking soda (sodium bicarbonate)** and **naturally carbonated mineral waters**. Physiologically, bicarbonate is primarily produced endogenously through the enzymatic action of

carbonic anhydrase, which converts **carbon dioxide (CO_2)** and **water (H_2O)** into **carbonic acid (H_2CO_3)**, and then dissociates into **H^+ and HCO_3^-**.

Bicarbonate serves as the **body's primary extracellular buffer**, helping to **maintain acid–base homeostasis** by neutralizing excess hydrogen ions (H^+). It plays a critical role in the **bicarbonate buffer system**, which works in conjunction with the **respiratory system** (via CO_2 regulation) and the **renal system** (via H^+ secretion and HCO_3^- reabsorption).

Low bicarbonate levels indicate **metabolic acidosis**, which can result from conditions such as **diabetic ketoacidosis, renal failure**, or **diarrhea**. This is typically accompanied by **compensatory hyperventilation** (Kussmaul breathing).

High bicarbonate levels may suggest **metabolic alkalosis**, often due to **vomiting, diuretic use**, or **excessive antacid ingestion**, or they may represent a **compensatory response to chronic respiratory acidosis**, as in **chronic obstructive pulmonary disease (COPD)**.

For nurses, monitoring arterial or venous bicarbonate levels is essential when interpreting **arterial blood gases (ABGs)**, evaluating acid–base status, and guiding appropriate fluid, electrolyte, and ventilatory management.

Figure 6.3 Natural Sources of Electrolytes

The following tables summarize the key electrolytes present in intracellular and extracellular fluids, highlighting their concentrations and primary physiological roles. Understanding these electrolytes is essential for maintaining fluid balance, nerve function, and overall cellular health.

Table 6.5 Major Electrolytes in Intracellular and Extracellular Fluids

Electrolyte / Ion	Primary Location	Main Functions	Clinical Significance	Dietary Sources
Sodium (Na^+)	Extracellular fluid	Regulates fluid balance, blood pressure, nerve signaling	Hyponatremia: confusion, seizures; Hypernatremia: dehydration, neurological symptoms	Table salt, processed foods, canned soups, cheese
Potassium (K^+)	Intracellular fluid	Maintains cardiac rhythm and muscle contraction	Hypokalemia: weakness, arrhythmia; Hyperkalemia: cardiac arrest	Bananas, oranges, potatoes, spinach, legumes
Calcium (Ca^{2+})	Bone, blood (ionized/bound)	Bone strength, muscle contraction, nerve function, clotting	Hypocalcemia: tetany, cramps; Hypercalcemia: stones, fatigue, confusion	Dairy products, leafy greens, fortified foods, sardines
Magnesium (Mg^{2+})	Intracellular fluid	Enzyme activation, neuromuscular stability	Hypomagnesemia: tremors, arrhythmia; Hypermagnesemia: hypotension, respiratory depression	Nuts, whole grains, green leafy vegetables, legumes

Table 6.6 Electrolytes in Bone and Extracellular Fluid and Their Roles

Electrolyte / Ion	**Primary Location**	**Main Functions**	**Clinical Significance**	**Dietary Sources**
Chloride (Cl^-)	Extracellular fluid	Osmotic pressure, acid–base balance	Hypochloremia: alkalosis, twitching; Hyperchloremia: acidosis, dehydration	Table salt, seaweed, tomatoes, olives, celery
Phosphate (PO_4^{3-})	Bone, intracellular fluid	ATP production, bone mineralization	Hypophosphatemia: weakness, bone pain; Hyperphosphatemia: kidney disease, calcification	Meat, fish, dairy, eggs, whole grains, legumes
Bicarbonate (HCO_3^-)	Extracellular fluid	Buffers blood pH	Low: metabolic acidosis; High: metabolic alkalosis or compensation	Baking soda (sodium bicarbonate), naturally carbonated mineral water (limited)

Case Study:

Ms. **Li Ming**, a 60-year-old woman, arrived at the emergency department complaining of painful muscle cramps in her legs, tingling in her fingers, and a pins-and-needles sensation around her mouth. On physical examination, she had a positive Trousseau's sign and exhibited mild muscle twitching. Her lab results revealed **hypocalcemia** with a calcium level of 7.1 mg/dL. She had a history of recent thyroidectomy and had not been taking calcium supplements post-surgery. Her diet was also low in dairy products and leafy greens.

Nursing Analysis and Intervention:

Ms. Li Ming's symptoms and lab findings are consistent with **hypocalcemia**, a common complication following **thyroid surgery**, often due to reduced secretion of **parathyroid hormone (PTH)**. According to Table 6.3, calcium is critical for **muscle contraction, nerve function, and blood clotting**, and its deficiency can lead to **tetany, cramps, and neuromuscular irritability**. The nurse's role includes monitoring for signs such as Chvostek's and Trousseau's signs, ensuring calcium levels are corrected either orally or intravenously as needed, and educating the patient on **calcium-rich dietary sources** like dairy, leafy greens, and sardines. Long-term management includes supplementation and follow-up to prevent recurrence.

Integration of Hormones, Electrolytes, and Vitamins in Physiological Regulation

The human body relies on a highly coordinated interplay between hormones, electrolytes, and vitamins to maintain internal balance and respond to physiological demands. These bioactive substances work synergistically through **feedback loops**, **cell signaling**, and **biochemical pathways** to regulate vital functions such as metabolism, fluid and electrolyte balance, bone health, immunity, and energy production.

Hormones serve as long-distance chemical messengers released by endocrine glands. They act on specific target tissues to regulate glucose metabolism (e.g., insulin and glucagon), stress responses (e.g., cortisol), fluid balance (e.g., ADH and aldosterone), and growth (e.g., GH and thyroid hormones). Their secretion is precisely controlled by **negative feedback mechanisms**, often involving the hypothalamic-pituitary axis and influenced by circulating levels of electrolytes and nutrients.

Electrolytes such as sodium, potassium, calcium, and bicarbonate are tightly regulated by both renal and hormonal systems. For example, **aldosterone** increases sodium reabsorption and potassium excretion to maintain blood pressure and fluid balance, while **parathyroid hormone (PTH)** regulates calcium and phosphate levels. Electrolyte shifts can influence hormone activity and vice versa, creating a dynamic balance that nurses must monitor closely in clinical practice.

Vitamins, particularly those involved as **coenzymes** (e.g., B-complex vitamins), are essential for the enzymatic reactions that drive metabolic pathways. Fat-soluble vitamins (A, D, E, K) have roles in gene expression, calcium absorption, antioxidant protection, and blood coagulation. Their function often intersects with hormonal and electrolyte regulation—such as **vitamin D** working with **PTH** to enhance calcium absorption and bone mineralization.

This integrated biochemical network ensures that any disturbance in one component—such as a hormonal disorder, electrolyte imbalance, or vitamin deficiency—can have cascading effects on other systems. Understanding these interconnections is essential for nurses to assess patient status holistically, interpret lab results accurately, and apply appropriate interventions in both acute and long-term care settings.

Chemistry Link to Health

Electrolyte–Endocrine Disruption Post-COVID-19

Emerging clinical studies underscore a critical connection between hormonal imbalances and electrolyte disturbances in patients recovering from COVID-19. Hyponatremia (low blood sodium levels) has been frequently reported in both acute and post-acute COVID-19 phases. This condition is often associated with dysregulation of hormones such as cortisol, aldosterone, antidiuretic hormone (ADH), and thyroid hormones, all of which are essential for maintaining fluid and sodium balance.

Recent evidence suggests that the SARS-CoV-2 virus may disrupt the renin-angiotensin-aldosterone system (RAAS) through its interaction with ACE2 receptors, impairing the body's ability to retain sodium and manage water homeostasis. Additionally, adrenal insufficiency or syndrome of inappropriate antidiuretic hormone secretion (SIADH) may further contribute to dilutional hyponatremia. These biochemical disturbances increase the risk of fatigue, confusion, orthostatic hypotension, and cardiovascular dysfunction—core symptoms commonly reported in patients experiencing long COVID.

Clinicians are urged to monitor electrolyte levels and hormonal status in post-COVID patients, particularly those presenting with nonspecific neurological or cardiac symptoms.

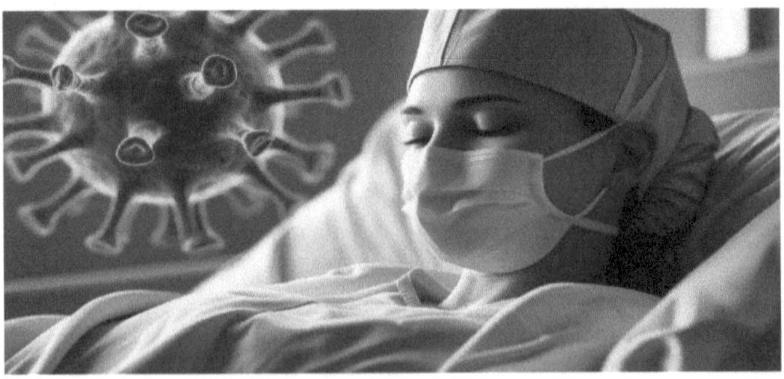

Chapter summary

Bioactive substances such as **vitamins**, **hormones**, and **electrolytes** are essential to maintaining human health and physiological stability.

- **Vitamins**, classified as either water-soluble (e.g., B-complex, C) or fat-soluble (e.g., A, D, E, K), function as cofactors in numerous metabolic processes. Deficiencies or excesses can lead to clinical conditions such as anemia, scurvy, rickets, or even toxicity and neurological disturbances.
- **Hormones** act as chemical messengers, regulating critical physiological processes including metabolism, growth, reproduction, and fluid-electrolyte balance. Dysregulation of hormones like insulin, cortisol, or thyroxine can manifest as diabetes mellitus, Cushing's syndrome, or hypothyroidism.
- **Electrolytes**—such as **sodium (Na^+)**, **potassium (K^+)**, **calcium (Ca^{2+})**, and **bicarbonate (HCO_3^-)**—are crucial for nerve impulse transmission, muscle contraction, fluid distribution, and acid–base homeostasis. Even minor disturbances in their concentrations may result in serious conditions like arrhythmias, seizures, muscle weakness, or coma.

For nursing professionals, a thorough understanding of these substances is essential for interpreting laboratory results, managing therapies, preventing complications, and delivering safe, evidence-based patient care.

EXERCISES

Multiple Choice

1. Which vitamin deficiency is associated with night blindness?
   A. Vitamin C B. Vitamin D C. Vitamin A D. Vitamin B12
2. Which hormone regulates blood calcium levels?
   A. Insulin B. Parathyroid hormone (PTH) C. Glucagon D. Aldosterone
3. Hypokalemia refers to low levels of:
   A. Calcium B. Sodium C. Potassium D. Magnesium
4. Which hormone increases blood glucose levels during fasting?
   A. Insulin B. ADH C. Glucagon D. Oxytocin

True or False

1. _____ Vitamin D deficiency can lead to rickets in children.
2. _____ Insulin raises blood sugar by promoting gluconeogenesis.
3. _____ ADH increases urine output by promoting diuresis.
4. _____ Hypocalcemia may cause muscle cramps and tetany.

Fill in the Blanks

1. The hormone responsible for lowering blood glucose is __________.
2. The electrolyte primarily responsible for maintaining fluid balance is __________.
3. A deficiency in vitamin B12 can result in __________ anemia.
4. Excessive secretion of aldosterone may cause __________ and __________.

Short Answer & Critical Thinking

1. Explain why a nurse should monitor potassium levels in a patient receiving loop diuretics.
2. Why is vitamin K routinely given to newborns?

Clinical Case Studies

Case 1: A 72-year-old male patient receiving IV fluids complains of muscle cramps and irregular heartbeat. Lab tests show low potassium levels.

- What is the likely condition?
- What nursing interventions are appropriate?

Case 2: A 30-year-old woman reports weight gain, cold intolerance, and fatigue. Blood tests reveal elevated TSH and low T4.

- What is the most likely diagnosis?
- What education should the nurse provide?

Chapter 7

Water, pH, and Fluid Balance

Chapter 7 :Water, pH, and Fluid Balance

Learning Objectives

By the end of this chapter, you should be able to:

1. Describe the key properties and functions of water in the human body.
2. Explain how water acts as a universal solvent in physiological processes.
3. Interpret the pH scale and mechanisms of acid–base balance.
4. Distinguish between acidosis and alkalosis with clinical relevance.
5. Describe fluid compartments and forces governing fluid movement.
6. Understand the chemical principles behind dialysis and its nursing applications.

Water is the foundation of life and the most abundant compound in the human body. It serves not only as a solvent and transport medium but also plays a critical role in regulating body temperature, facilitating biochemical reactions, and maintaining cellular integrity. For nurses, understanding the chemical properties of water and its interactions within the body is essential for managing hydration, administering intravenous fluids, interpreting acid–base disorders, and supporting organ function.

This chapter explores the unique characteristics of water, the principles of pH and acid–base balance, and the mechanisms governing fluid movement across body compartments. It also examines the hormonal regulation of fluid balance and the clinical application of dialysis as a life-saving therapy in kidney failure. Through a chemistry-based lens, students will gain insights into the physiological processes that underlie hydration, electrolyte homeostasis, and acid–base stability—core components of safe and effective nursing care.

Water: Properties and Biological Significance

Water is a polar molecule, characterized by a partial positive charge on its hydrogen atoms and a partial negative charge on its oxygen atom. This polarity enables water molecules to form hydrogen bonds, which in turn make water an exceptional solvent for ions, polar compounds, and biologically important molecules. Within the human body, water plays several vital roles: it facilitates the transport of nutrients, gases, and waste products; helps regulate body temperature through perspiration and blood circulation; and provides essential lubrication in joints and soft tissues. Additionally, water's high specific heat capacity allows it to absorb and store heat with minimal temperature change, thereby ensuring thermal stability at the cellular and organ levels.

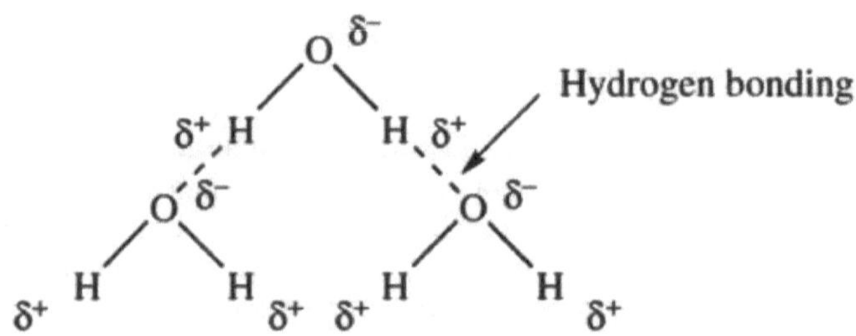

Figure 7.1 Hydrogen bonding

Water as a Universal Solvent

Water is commonly known as the universal solvent due to its remarkable ability to dissolve a vast array of substances, particularly ionic compounds and polar molecules. This capacity arises from water's molecular polarity and its tendency to form hydrogen bonds with solutes. In the human body, this property is critical for dissolving electrolytes, nutrients, gases (such as oxygen and carbon dioxide), hormones, and metabolic waste. Consequently, nearly all biochemical reactions occur in aqueous environments, with water acting as the primary medium for transporting substances through blood, lymph, and intracellular fluids. For nursing professionals, understanding water's solvent role is essential when preparing intravenous (IV) fluids, administering medications, and managing hydration and electrolyte therapy, as poor solubility may lead to drug incompatibility, precipitation, or diminished therapeutic effectiveness.

Unique Properties of Water

Water possesses a set of unique physical and chemical properties that are vital for sustaining life and supporting physiological functions. These properties not only contribute to the biochemical stability of the human body but also play critical roles in fluid regulation, temperature control, and solute transport. The table below outlines the most significant characteristics of water and their clinical and physiological relevance, especially in nursing practice.

Table 7.1 Important characteristics of water

Property	Scientific Explanation	Clinical/Physiological Relevance
Polarity	Water's partial charges dissolve ionic/polar substances	Aids transport of electrolytes, glucose, medications
Universal Solvent	Dissolves more substances due to polarity and H-bonds	Facilitates biochemical reactions, drug solubility
High Specific Heat	Absorbs heat with minimal temperature change	Maintains body temperature, buffers thermal shifts
High Heat of Vaporization	Needs much energy to vaporize	Enables cooling via sweating during fever/activity
Cohesion and Adhesion	Molecules attract each other and surfaces	Supports capillary action and fluid movement
Lower Density of Ice	Ice less dense than water, floats	Insulates aquatic ecosystems (indirect clinical)

Hydrolysis: A Key Reaction in Physiology

Hydrolysis is a fundamental chemical reaction in which a molecule of water is used to cleave a chemical bond within a larger compound, producing two smaller molecules. The word "hydrolysis" literally means "splitting by water." This reaction is essential for numerous physiological functions, especially in digestion, metabolism, and cellular regulation. For instance, digestive enzymes—such as amylase, protease, and lipase—catalyze the hydrolysis of carbohydrates, proteins, and lipids into their absorbable components: glucose, amino acids, and fatty acids. In nucleic acid metabolism, hydrolysis breaks the bonds between nucleotides in DNA and RNA. Clinically, hydrolysis is central to drug metabolism, where water-dependent enzymes convert medications into active or excretable forms. For nurses, understanding hydrolysis is vital when addressing enzymatic deficiencies, malabsorption syndromes, or administering prodrugs that require activation through hydrolytic reactions.

$$\underset{\text{Reactant}}{A-B} \quad + \quad \underset{\text{Water}}{H_2O} \longrightarrow \underset{\text{Product 1}}{A-H} \quad + \quad \underset{\text{Product 2}}{B-OH}$$

Figure 7.2 Hydrolysis general reaction

pH and Acid–Base Balance

The pH of a solution indicates its hydrogen ion concentration [H^+], providing a measure of its acidity or alkalinity. The pH scale ranges from 0 to 14, with 7 representing a neutral solution. In human physiology, arterial blood pH is precisely regulated between 7.35 and 7.45 to ensure optimal enzyme activity, cellular metabolism, and structural stability of proteins. A pH below 7.35 indicates **acidosis**, while a pH above 7.45 signifies **alkalosis**. These disturbances may arise from **respiratory causes**—such as altered carbon dioxide (CO_2) levels—or **metabolic causes**, including changes in bicarbonate (HCO_3^-) concentration or the buildup of organic acids.

To maintain pH homeostasis, the body relies on three key mechanisms: **buffer systems**, **respiratory regulation**, and **renal compensation**. The **bicarbonate buffer system** (HCO_3^-/H_2CO_3) is the most critical, as it rapidly neutralizes excess hydrogen or hydroxide ions. The **lungs** regulate pH by adjusting the exhalation of

CO_2, while the **kidneys** contribute to long-term balance by excreting H^+ and reabsorbing bicarbonate. For nurses, a deep understanding of acid–base regulation is essential for interpreting **arterial blood gases (ABGs)**, identifying disorders such as **diabetic ketoacidosis**, **respiratory failure**, or **metabolic alkalosis**, and making informed decisions regarding **IV fluid therapy** and **ventilator management**.

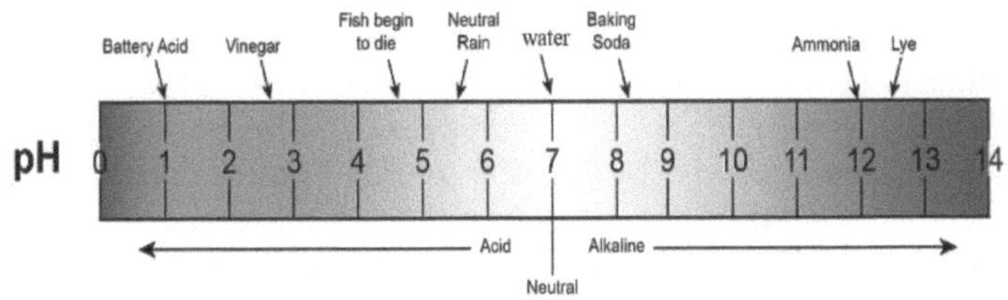

Figure 7.3 The pH scale

Acidosis occurs when the blood pH drops below **7.35**, indicating an increase in acidity. It can result from **respiratory causes**, such as hypoventilation or respiratory failure, which lead to elevated carbon dioxide (↑ CO_2) levels; or from **metabolic causes**, including the accumulation of organic acids (e.g., ↑ **lactic acid**) or a reduction in bicarbonate (↓ HCO_3^-), as seen in renal failure or diabetic ketoacidosis.

Conversely, **alkalosis** occurs when blood pH exceeds **7.45**, indicating increased basicity. This may be caused by **respiratory alkalosis**, such as **hyperventilation** that lowers carbon dioxide levels (↓ CO_2), or **metabolic alkalosis**, often due to excessive bicarbonate intake or loss of stomach acid through prolonged vomiting. Both conditions disrupt cellular homeostasis and require prompt nursing assessment and intervention based on **arterial blood gas (ABG)** analysis and clinical signs.

Case Study: Diabetic Ketoacidosis (DKA)

A 24-year-old female patient with type 1 diabetes is brought to the emergency department with complaints of nausea, vomiting, abdominal pain, and deep, rapid breathing. The nurse notes a fruity odor on her breath and observes signs of dehydration and altered mental status. Vital signs reveal a respiratory rate of 30 breaths per minute, heart rate of 118 bpm, blood pressure of 95/60 mmHg, and a random blood glucose level of 480 mg/dL.

Arterial blood gas (ABG) analysis shows a pH of 7.20, a low bicarbonate level (HCO_3^- = 10 mEq/L), and a decreased partial pressure of carbon dioxide ($PaCO_2$ = 25 mmHg). These findings are consistent with **metabolic acidosis** due to the accumulation of ketone bodies in the blood—a hallmark of diabetic ketoacidosis.

Nursing assessment confirms that the patient's hyperventilation (Kussmaul respirations) is a compensatory response aimed at lowering blood acidity by expelling CO_2. The presence of low bicarbonate and acidic pH indicates a primary metabolic cause, while the decreased CO_2 reflects partial respiratory compensation.

Immediate nursing interventions include administering intravenous normal saline (0.9% NaCl) to restore fluid volume, initiating an insulin drip to reduce blood glucose and halt ketone production, and monitoring blood glucose levels, serum electrolytes (particularly potassium), and urine output. ABG values are reassessed regularly to evaluate the effectiveness of treatment and ensure stabilization of the patient's acid–base status.

This case highlights a critical example of metabolic acidosis and the importance of prompt recognition and management. For nurses, understanding the principles of acid–base balance, interpreting ABGs, and identifying compensatory mechanisms are essential skills in emergency and critical care settings.

Buffer Systems: The Body’s First Line of pH Defense

The human body constantly produces acids and bases as byproducts of metabolism. Without regulation, these substances could dramatically alter blood pH and disrupt cellular function. Buffer systems act as the body’s first line of defense, minimizing changes in pH by neutralizing excess hydrogen (H^+) or hydroxide (OH^-) ions. Three primary buffer systems operate in the body: the bicarbonate buffer, phosphate buffer, and protein buffer systems—each functioning in distinct compartments and playing critical roles in maintaining acid–base homeostasis.

1. Bicarbonate Buffer System

The bicarbonate buffer system (HCO_3^-/H_2CO_3) is the most important buffer in extracellular fluid and blood plasma. It relies on the reversible reaction between carbonic acid (H_2CO_3) and bicarbonate ions (HCO_3^-) to regulate hydrogen ion concentration. When acidity increases, bicarbonate binds with excess H^+ to form carbonic acid, which then decomposes into water and carbon dioxide (CO_2), the latter being exhaled by the lungs. Conversely, in basic conditions, carbonic acid dissociates to release H^+ and neutralize excess hydroxide. This system is regulated by the lungs—through CO_2 removal—and the kidneys—through H^+ excretion and bicarbonate reabsorption—making it central to ABG interpretation and clinical acid–base management.

2. Phosphate Buffer System

The phosphate buffer system ($H_2PO_4^-/HPO_4^{2-}$) is more effective in the intracellular environment and the renal tubules. It operates by converting dihydrogen phosphate ($H_2PO_4^-$) into monohydrogen phosphate (HPO_4^{2-}) and vice versa, depending on the H^+ concentration. It plays a vital role in urine pH regulation and acid excretion in the kidneys. While it contributes less to buffering blood plasma, it is essential for maintaining cellular pH balance and for renal function.

3. Protein Buffer System

Protein buffer systems—especially those involving hemoglobin—are crucial in both intracellular and extracellular fluids. Amino acid side chains in proteins can accept or donate H^+, enabling them to buffer changes in pH. Hemoglobin in red blood cells binds H^+ ions generated during CO_2 transport, preventing acidosis. Plasma proteins such as albumin also help maintain pH in the extracellular fluid. The protein buffer system is particularly important during respiratory or metabolic disturbances, making it a critical element in clinical acid–base balance.

Table 7.2 Buffer Systems in the Human Body

Buffer System	Components	Primary Location	Clinical Importance
Bicarbonate Buffer System	H_2CO_3 / HCO_3^-	Extracellular fluid, blood plasma	Primary blood buffer; regulated by lungs and kidneys; critical for interpreting ABGs
Phosphate Buffer System	$H_2PO_4^-$ / HPO_4^{2-}	Intracellular fluid, renal tubules	Important in kidney function and urine pH regulation
Protein Buffer System	Amino acid side chains (e.g., hemoglobin)	Blood cells, plasma, intracellular fluid	Buffers H^+ in tissues and blood; hemoglobin buffers during CO_2 transport

Fluid Balance in the Body

Fluid balance refers to the maintenance of the proper amount and distribution of water within the body's compartments to ensure normal physiological function. Total body water is divided between **intracellular fluid (ICF)**—which makes up approximately two-thirds of total body water—and **extracellular fluid (ECF)**, which includes plasma and interstitial fluid. Water moves between these compartments by **osmosis**, driven by solute concentrations, particularly electrolytes like sodium and potassium. Fluid balance is regulated by complex hormonal mechanisms involving **antidiuretic hormone (ADH)**, which increases water reabsorption in the kidneys; **aldosterone**, which promotes sodium and water retention; and **atrial natriuretic peptide (ANP)**, which encourages sodium and water excretion. Disruption in this balance may lead to **dehydration** (fluid deficit), **overhydration** (fluid excess), or **edema** (fluid accumulation in tissues), each of which can significantly impact cardiovascular, renal, and neurological function. For nurses, assessing fluid status through **intake and output monitoring**, **daily weights**, **vital signs**, and **lab values (e.g., sodium, hematocrit, urine specific gravity)** is essential for preventing complications and guiding fluid therapy interventions.

Table 7.3 Fluid Compartments and Body Water Distribution

Fluid Compartment	Approx. % of Total Body Water	Location	Function
Intracellular Fluid (ICF)	≈ 66%	Inside cells	Supports cellular metabolism, enzyme activity, and chemical reactions
Extracellular Fluid (ECF)	≈ 33%	Outside cells	Divided into plasma and interstitial fluid; transports nutrients, hormones, and waste
Plasma	~7%	In blood vessels	Maintains blood pressure and transports blood cells
Interstitial Fluid	~26%	Between cells	Provides medium for exchange of gases, nutrients, and wastes between blood and tissues

Fluid movement between compartments is governed by: Osmosis, Hydrostatic pressure and Oncotic pressure.

Osmosis is the passive movement of **water molecules** across a **semipermeable membrane** from an area of **lower solute concentration** to an area of **higher solute concentration**. It is a fundamental process in maintaining **fluid balance** between the **intracellular and extracellular compartments** of the body. Unlike diffusion, which involves the movement of solutes, osmosis exclusively involves water. This process is driven by **osmotic pressure**, and continues until equilibrium is reached. In clinical settings, osmosis explains why **cells shrink (crenate)** in hypertonic solutions or **swell and possibly burst** in hypotonic environments. It also underlies the mechanism of **IV fluid therapy**: for instance, isotonic solutions maintain equilibrium, while hypertonic solutions draw fluid out of cells, and hypotonic solutions rehydrate cells. For nurses, understanding osmosis is essential when managing **dehydration, edema, electrolyte disturbances**, and **administering appropriate IV fluids** based on a patient's specific condition.

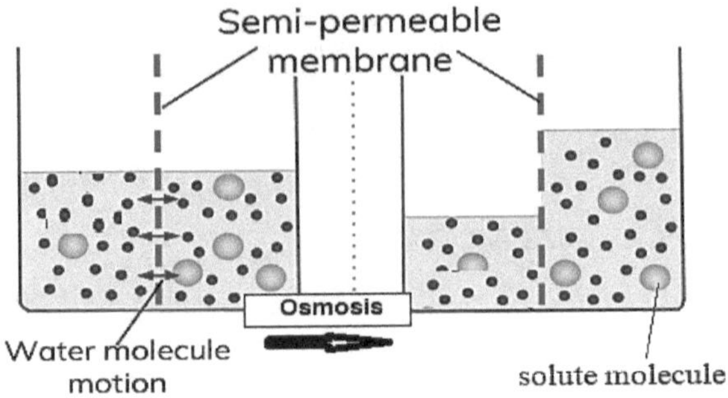

Figure 7.4 Osmosis pressure

Types of IV Solutions Based on Tonicity and Their Clinical Applications

Intravenous fluids are classified into three main categories based on tonicity—**isotonic**, **hypotonic**, and **hypertonic**—which refers to the concentration of solutes in the solution relative to that inside the body's cells. Tonicity determines the direction of water movement across the semipermeable cell membrane, affecting cellular volume, fluid distribution, and patient outcomes. An understanding of these distinctions is essential for nurses when selecting IV fluids appropriate to clinical conditions such as dehydration, electrolyte imbalances, or neurological concerns.

Isotonic solutions have a solute concentration that closely matches that of blood plasma (approximately 275–295 mOsm/L), resulting in no net movement of water across the cell membrane. This maintains normal cell shape and volume, making isotonic fluids ideal for restoring circulating volume, treating mild dehydration, managing postoperative fluid losses, or stabilizing patients with mild hyponatremia. Common examples include **0.9% sodium chloride**

(Normal Saline), which contains 154 mEq/L of sodium and chloride and is widely used for general fluid replacement. **Lactated Ringer's solution**, which contains sodium, potassium, calcium, chloride, and lactate, is commonly used in surgical, burn, or trauma cases. Additionally, **5% dextrose in water (D5W)** is isotonic in the IV bag but becomes hypotonic in the body after glucose is metabolized.

In contrast, **hypotonic solutions** contain a lower concentration of solutes than intracellular fluid, causing water to shift into cells, leading to cellular swelling and potential lysis if overused. These fluids are indicated in conditions involving cellular dehydration, such as **diabetic ketoacidosis (DKA)** or **hypernatremia**, as they help rehydrate the intracellular compartment. Examples include **0.45% sodium chloride (half-normal saline)** and **0.33% sodium chloride**, both of which must be used with caution to avoid complications like **cerebral edema**, particularly in patients with head injuries or neurological conditions. D5W also functions as a hypotonic fluid once the glucose is metabolized, making it useful in providing hydration and energy but potentially hazardous in fluid-restricted or brain-injured patients.

Hypertonic solutions, on the other hand, have a higher solute concentration than plasma, drawing water out of cells and causing them to shrink (crenate). These fluids are used in situations such as **severe hyponatremia**, **cerebral edema**, or **increased intracranial pressure**, where reducing intracellular fluid is clinically necessary. Notable examples include **3% sodium chloride**, which requires careful administration, usually through a central venous line due to its high osmolarity and risk of vein irritation. Other hypertonic fluids such as **D5NS (5% dextrose in normal saline)** and **D10W (10% dextrose in water)** are used in critical care to provide both sodium and energy or to support patients receiving parenteral nutrition. Despite their benefits, hypertonic solutions pose risks including **cellular dehydration**, **volume overload**, and **hypernatremia**, necessitating close monitoring of serum electrolytes, urine output, lung sounds, and cardiovascular status.

In summary, selecting the appropriate IV fluid depends on a thorough assessment of the patient's clinical condition, electrolyte status, fluid volume, and organ function. A clear understanding of how isotonic, hypotonic, and hypertonic solutions affect fluid compartments and cellular physiology enables nurses to safely manage fluid therapy and prevent serious complications associated with inappropriate fluid administration.

Case Study: Hypotonic IV Fluid in Hypernatremia

A 65-year-old male presents with confusion, dry mucous membranes, and a history of diarrhea and poor fluid intake. Labs reveal hypernatremia (serum sodium = 155 mEq/L) and signs of mild dehydration. His blood pressure is 98/62 mmHg, and his respiratory rate is elevated.

The physician orders **0.45% sodium chloride (half-normal saline)** to correct intracellular dehydration and gradually reduce sodium levels. This hypotonic fluid promotes water movement into cells, rehydrating tissues.

The nurse initiates the IV infusion slowly and closely monitors the patient's neurological status, urine output, and serum electrolytes to prevent complications such as cerebral edema. Over several hours, the patient's mental status improves, and sodium levels decline gradually.

This case highlights the importance of understanding IV fluid tonicity. While the **physician selects the appropriate therapy**, the **nurse plays a vital role** in monitoring treatment response and ensuring patient safety.

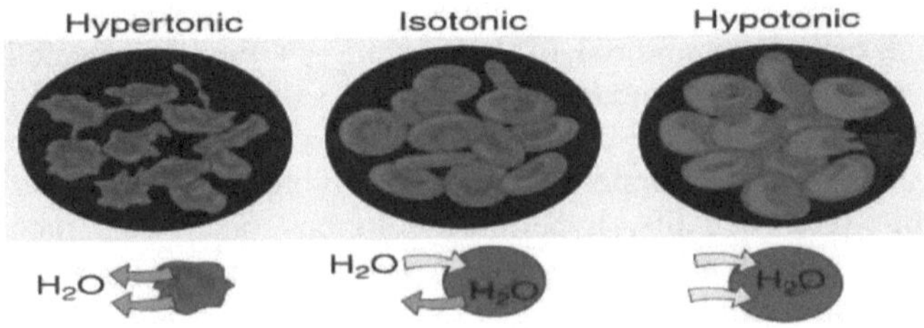

Figure 7.5 (a) Crenation: In a hypertonic solution, water leaves the red blood cell, causing it to shrink. (b) In an isotonic solution, a red blood cell retains its normal volume , (c) Hemolysis: In a hypotonic solution, water flows into a red blood cell, causing it to swell and burst.

Table7.4 Comparison of IV Solutions Based on Tonicity

Type of Solution	Relative Solute Concentration	Effect on Cells	Common Examples	Clinical Applications
Isotonic	Same as plasma (≈ 275–295 mOsm/L)	No net water movement; cells retain shape	0.9% NaCl (Normal Saline), Lactated Ringer's, D5W (initially)	Fluid resuscitation, mild dehydration, post-op fluid replacement
Hypotonic	Lower than plasma	Water enters cells → swelling or lysis	0.45% NaCl, 0.33% NaCl, D5W (after metabolism)	Cellular dehydration (e.g., DKA, hypernatremia); caution in neurological patients
Hypertonic	Higher than plasma	Water leaves cells → shrinkage (crenation)	3% NaCl, D5NS, D10W	Severe hyponatremia, cerebral edema, energy/ nutrition supplementation in critical care.

Hydrostatic Pressure

Hydrostatic pressure is the force exerted by a fluid against the walls of its container. In the human body, it refers mainly to the pressure exerted by blood on the walls of capillaries, pushing plasma and small solutes through the capillary walls into the surrounding interstitial tissue. This outward pressure plays a key role in facilitating the exchange of oxygen, nutrients, and waste products between the blood and body tissues. Hydrostatic pressure is typically highest at the arterial end of the capillary and gradually decreases toward the venous end, where oncotic pressure—generated by plasma proteins like albumin—draws fluid back into the bloodstream. An imbalance between these forces can lead to pathological fluid accumulation.

For example, elevated hydrostatic pressure due to conditions such as congestive heart failure, deep vein thrombosis, or fluid overload can result in edema. For nurses, recognizing and managing changes in hydrostatic pressure is critical in fluid therapy, IV management, and caring for patients with cardiovascular or renal conditions.

Figure 7.6 Hydrostatic Pressure

Oncotic Pressure

Oncotic pressure, also known as colloid osmotic pressure, is the component of osmotic pressure exerted by plasma proteins—primarily albumin—within the blood vessels. It plays a critical role in maintaining fluid balance between the intravascular (plasma) and interstitial compartments. While hydrostatic pressure drives fluid out of the capillaries, oncotic pressure counteracts this by drawing fluid back into the capillaries, especially at the venular end of the capillary bed. This dynamic interaction between hydrostatic and oncotic forces is governed by Starling's law of capillary exchange. A decrease in oncotic pressure, commonly resulting from hypoalbuminemia associated with liver disease, nephrotic syndrome, or severe malnutrition, can lead to edema due to excessive fluid accumulation in the interstitial space. Conversely, restoring oncotic pressure through the administration of colloid intravenous solutions—such as albumin infusions—can help reestablish vascular volume by shifting fluid back into the circulation. For nurses, assessing oncotic pressure indirectly via serum albumin levels and managing its effects is essential in fluid therapy, particularly for critically ill or volume-depleted patients.

The movement of fluids within the body is governed by several key physical forces. Table 7.4 summarizes and compares the roles of osmosis, hydrostatic pressure, and oncotic pressure in maintaining fluid balance.

Table 7.4 Comparison of Osmosis, Hydrostatic Pressure, and Oncotic Pressure

Force	Mechanism	Physiological Role
Osmosis	Movement of water across a semipermeable membrane from low to high solute concentration	Maintains cellular hydration and balance between intracellular and extracellular compartments
Hydrostatic Pressure	Force exerted by fluid against vessel walls	Pushes fluid out of capillaries into interstitial space; promotes filtration
Oncotic Pressure	Osmotic pressure generated by plasma proteins (e.g., albumin)	Pulls fluid back into capillaries; opposes hydrostatic pressure and prevents edema

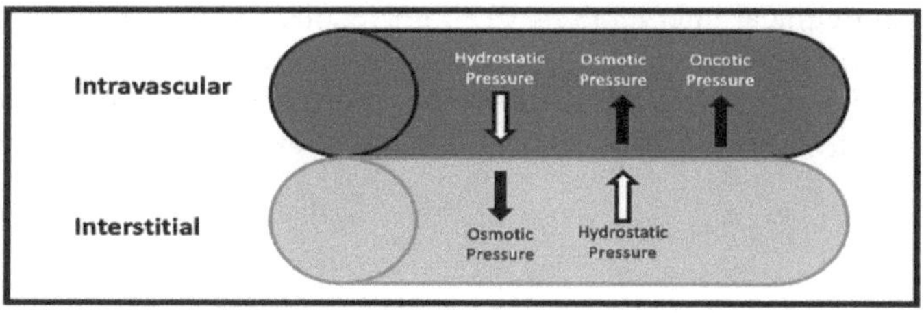

Figure 7.7 Starling Forces

Hormonal Regulation of Fluid Balance

The regulation of fluid and electrolyte balance in the human body is primarily controlled by hormonal mechanisms that coordinate water and sodium homeostasis across different fluid compartments. Three major hormones are involved in this process. Antidiuretic hormone (ADH), also known as vasopressin, is secreted by the posterior pituitary gland and acts on the kidneys to increase water reabsorption in the collecting ducts. This concentrates the urine and conserves body water, helping to prevent dehydration. Aldosterone, a steroid hormone produced by the adrenal cortex, promotes sodium reabsorption in the renal tubules; since water follows sodium osmotically, this leads to an increase in blood volume and pressure. In contrast, atrial natriuretic peptide (ANP) is released by atrial myocytes in response to atrial stretching from increased blood volume. ANP enhances the excretion of sodium and water, thus reducing blood volume and pressure.

These hormones work in a finely tuned balance to maintain optimal blood pressure and fluid volume, adapting to changes such as dehydration, hemorrhage, or fluid overload. The feedback mechanisms involved rely on sensors like baroreceptors and osmoreceptors that detect shifts in blood pressure and plasma osmolality. Dysregulation of these hormonal systems can lead to fluid imbalance conditions such as dehydration, fluid overload, edema, and electrolyte disturbances. For nurses, a clear understanding of these hormonal pathways is essential for evaluating a patient’s hydration status, interpreting serum electrolyte levels, and managing clinical interventions like intravenous fluid administration and diuretic therapy.

Table 7.5 summarizes these three key hormones, outlining their sources, physiological actions, and clinical relevance in maintaining fluid balance.

Table 7.5 Hormonal Regulation of Fluid Balance

Hormone	Source	Main Action	Clinical Significance
ADH (Antidiuretic Hormone)	Posterior pituitary	Increases water reabsorption in kidneys	Conserves water; low levels cause diabetes insipidus; high levels contribute to SIADH
Aldosterone	Adrenal cortex	Promotes sodium (and water) reabsorption; potassium excretion	Maintains blood pressure; excess causes fluid retention and hypokalemia
ANP (Atrial Natriuretic Peptide)	Cardiac atria	Promotes sodium and water excretion	Reduces blood volume and pressure; counteracts effects of ADH and aldosterone

Example Case Study: Hormonal Imbalance and Nursing Management

Case: A 72-year-old male patient is admitted with congestive heart failure. He presents with shortness of breath, swollen ankles, elevated blood pressure, and decreased urine output.

Nursing Assessment: Lab results show elevated BNP (similar to ANP), low sodium levels, and signs of fluid overload on chest X-ray. The patient is retaining water due to overactivation of ADH and aldosterone.

Intervention: The nurse initiates fluid restriction, administers loop diuretics as prescribed, and monitors electrolyte levels, weight, and urine output. Understanding the hormonal dysregulation helps guide clinical decisions and prevent complications such as hyponatremia or hypokalemia.

Dialysis: Clinical Chemistry in Action

Dialysis is a critical life-sustaining therapy for patients with advanced kidney failure. When the kidneys can no longer filter waste, balance fluids, or regulate electrolytes and pH, dialysis steps in to perform these vital functions. At its core, dialysis is a direct application of fundamental chemical principles—especially diffusion, osmosis, and selective membrane transport. Understanding these concepts allows nurses to anticipate how solutes and fluids move during treatment, and to monitor the patient for signs of imbalance or complication.

In hemodialysis, the patient's blood is pumped through an external filter called a dialyzer, which contains a semipermeable membrane. On one side of the membrane is the patient's blood; on the other side is a carefully prepared chemical solution known as the dialysate. Small waste molecules like urea, creatinine, and excess potassium diffuse from the blood (where they are in high concentration) into the dialysate (which contains none or very little), following the concentration gradient. At the same time, excess water may be drawn out through osmotic movement, depending on the solute concentration of the dialysate. Larger molecules such as proteins and blood cells are too big to cross the membrane and are retained in the bloodstream.

The dialysate solution itself must be chemically balanced to mimic normal plasma composition. It typically contains sodium, chloride, bicarbonate, calcium, magnesium, and sometimes glucose. The exact concentration of each component is adjusted based on the patient's latest lab values and clinical condition. For example, a patient with hyperkalemia will receive a dialysate with little or no potassium to promote potassium removal. In patients with metabolic acidosis, bicarbonate in the dialysate acts as a buffer to restore normal blood pH. These chemical adjustments require precision, and nurses must understand how even small variations in dialysate composition can affect the patient's safety and recovery.

Peritoneal dialysis, another form of therapy, uses the patient's own peritoneal membrane as the filtering surface. Dialysate is infused into the abdominal cavity and left in place for several hours, allowing solute exchange across the peritoneal capillaries. The same chemical principles apply—waste products and excess electrolytes move into the dialysate by diffusion, while fluid is removed via osmosis. Although peritoneal dialysis is slower and less efficient per session than hemodialysis, it is gentler and can be done at home, making it suitable for certain patients.

From a nursing perspective, dialysis care requires more than following procedures—it demands chemical thinking. Nurses must monitor vital signs, fluid balance, weight changes, and laboratory values such as potassium, calcium, bicarbonate, and pH before and after treatment. They must recognize symptoms of dialysis-related complications like hypotension, cramping, or disequilibrium syndrome, which occur due to rapid shifts in solute concentration. A strong grasp of the chemistry involved in dialysis enables nurses to respond proactively, communicate effectively with the dialysis team, and advocate for safe, individualized patient care.

Ultimately, dialysis exemplifies how chemistry operates not only at the molecular level but also at the heart of clinical decision-making. It transforms abstract concepts like diffusion and concentration gradients into real-life interventions that save lives and restore balance in the body. For nurses, mastering these principles is key to delivering confident, competent, and compassionate care to patients undergoing this essential therapy.

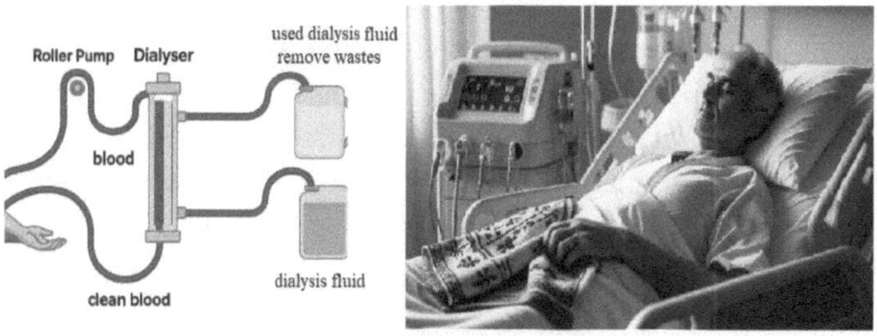

Figure 7.8 Dialysis

Chemistry Link to Health

Hydration and Cognitive Function in Older Adults

Recent research highlights the critical role of hydration in preserving cognitive function, particularly among older adults. A 2023 prospective study published in BMC Medicine followed nearly 900 elderly individuals with metabolic syndrome or obesity over a two-year period. The study found that participants with physiological dehydration—defined as serum osmolarity ≥ 300 mmol/L—exhibited a significant decline in global cognitive performance compared to well-hydrated individuals. Notably, for every 1 mmol/L increase in serum osmolarity, there was a measurable decline in cognitive function ($\beta = -0.010$, $p = 0.002$), suggesting even mild dehydration can have a meaningful impact on brain health.

Further evidence from a 2024 study in Scientific Reports demonstrated that changes in total body water and blood volume can influence vascular resistance. This, in turn, may contribute to symptoms such as dizziness and chronic pain through disrupted perfusion and altered fluid dynamics.

Together, these findings emphasize the need for proactive hydration assessment in older adults—not merely by tracking fluid intake, but by monitoring markers such as serum osmolarity. Maintaining proper fluid balance within both intracellular and intravascular compartments is essential not only for cognitive preservation but also for overall physiological stability and patient well-being.

Chapter Summary

Water is the most abundant and vital compound in the human body, playing a central role in sustaining life and supporting all physiological functions. It acts as a universal solvent, facilitates chemical reactions, regulates body temperature, and maintains cellular integrity. Acid–base homeostasis is preserved within a narrow blood pH range of 7.35 to 7.45, primarily through buffer systems, respiratory adjustments, and renal mechanisms. Fluid balance is maintained by the distribution of water between intracellular and extracellular compartments, influenced by osmotic gradients, hydrostatic pressure, and oncotic pressure. Hormones such as antidiuretic hormone (ADH), aldosterone, and atrial natriuretic peptide (ANP) coordinate the regulation of water and electrolyte levels. Disruptions in these regulatory systems can lead to dehydration, fluid overload, acid–base disturbances, or edema. For nurses, understanding these core principles is essential for effective clinical assessment, fluid therapy management, and patient safety.

EXERCISES

Multiple Choice

1. Which of the following is considered an isotonic solution?

   A. 0.9% NaCl B. 0.45% NaCl C. 3% NaCl D. 5% Dextrose in water

2. Which hormone promotes sodium and water retention in the kidneys?

   A. ADH B. Cortisol C. ANP D. Aldosterone

3. What does a blood pH of 7.50 indicate?

   A. Acidosis B. Alkalosis C. Normal D. Neutral

4. Which pressure is responsible for pulling fluid into the capillaries?

   A. Hydrostatic pressure B. Oncotic pressure
   C. Atmospheric pressure D. Osmotic pressure

5. The primary site of intracellular fluid is:

   A. Blood plasma B. Lymph
   C. Inside cells D. Interstitial fluid

True or False

1. Water moves from areas of low solute concentration to high solute concentration through osmosis.
2. Hypernatremia means there is too little sodium in the blood.
3. The normal range for blood pH is 7.0 to 7.3.
4. ADH increases water excretion from the kidneys.
5. Albumin contributes to oncotic pressure in capillaries.

Fill in the Blanks

1. The normal range of arterial blood pH is __________ to __________.
2. The force that pushes fluid out of the capillaries is called __________ pressure.
3. The hormone that promotes water reabsorption in the kidneys is __________.
4. __________ is the movement of water through a semipermeable membrane.
5. __________ pressure pulls fluid back into the blood vessels.

Critical Thinking Questions

1. Explain the importance of buffer systems in maintaining acid–base balance in the body.
2. Describe how dehydration affects cellular function and overall fluid balance.
3. Discuss how a nurse should assess fluid status in a patient who is vomiting and experiencing diarrhea.

Clinical Case Scenarios

1. Case 1: A 65-year-old male is admitted with signs of confusion, dry mucous membranes, and low blood pressure. His labs show elevated sodium and hematocrit. - What condition is likely? What nursing interventions would be appropriate?

2. Case 2: A patient receiving IV fluids develops shortness of breath and swelling in the lower limbs. You notice decreased serum sodium and increased blood pressure. - What condition might this indicate? What should be the next steps in management?

3. Case 3: A patient has a pH of 7.28 and elevated CO_2 levels on an ABG. He is lethargic and has a history of COPD. - What type of acid–base imbalance is this? How should the care plan be adjusted?

Chapter 8

Solutions in Clinical Practice

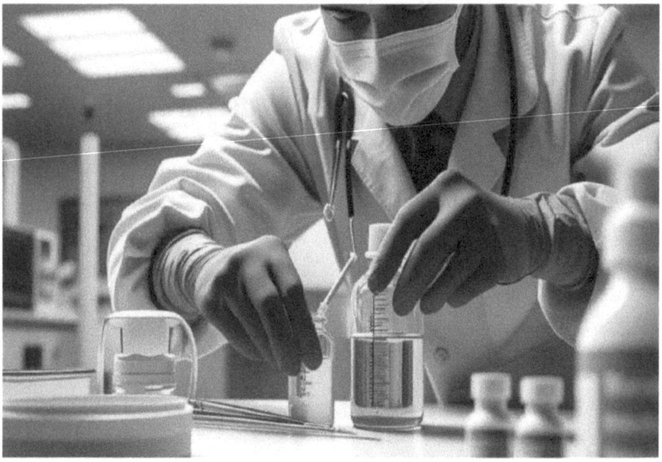

Chapter 8: Solutions in Clinical Practice

Learning Objectives

By the end of this chapter, you should be able to:

1. Identify types of solutions and their clinical uses.
2. Use concentration units (% w/v, % v/v, molarity, mEq/L) in calculations.
3. Calculate correct dilutions and drug dosages.
4. Prepare solutions using safe and sterile techniques.
5. Recognize IV fluid and drug compatibility issues.
6. Interpret lab values to guide fluid therapy.

Solutions are fundamental to modern healthcare, forming the chemical foundation for a broad range of essential procedures—from intravenous (IV) fluid preparation and medication reconstitution to precise drug dosing and electrolyte management. In clinical settings, solutions are more than mere mixtures of substances; they are life-sustaining interventions that demand accurate formulation and careful administration tailored to each patient's physiological status. This chapter explores the chemistry of solutions and their direct application in healthcare. Readers will learn to differentiate between true solutions, colloids, and suspensions, and to utilize clinical concentration units such as molarity, percent solutions, and normality with confidence. The chapter also emphasizes practical skills in dilution and dosage calculations, preparation of common medical solutions such as saline and dextrose, and adherence to sterility and compatibility standards. Mastering these concepts enables healthcare professionals to ensure safe and effective medication administration, minimize clinical errors, and deliver high-quality patient care. Although solution preparation may appear routine, it requires a solid understanding of chemistry and an unwavering commitment to precision—core competencies that this chapter seeks to instill.

What Is a Solution?

A **solution** is a homogeneous mixture formed when one substance (the **solute**) dissolves uniformly into another (the **solvent**), creating a single-phase system. In healthcare, solutions are essential for delivering medications, rehydrating patients, maintaining electrolyte balance, and supporting nutrition.

Key Components:

- **Solute**: The substance that is dissolved (e.g., salt, glucose, medication).
- **Solvent**: The substance that dissolves the solute—most often **water** in clinical settings.

Characteristics of Solutions:

Solutions exhibit several key characteristics that make them indispensable in clinical practice. Firstly, they are **homogeneous mixtures**, meaning the solute is evenly distributed throughout the solvent—whether in an IV saline drip or an oral medication, every portion has a consistent solute-to-solvent ratio. This uniformity ensures **reliable therapeutic effects** across all doses.

Secondly, solutions are typically **stable**: the solute remains dissolved and does not precipitate over time, a property that is vital for **safe storage and precise administration**. In **true solutions** (such as glucose or electrolyte solutions), the mixture is **optically clear**, allowing light to pass through without scattering. This distinguishes them from colloids or suspensions, which often appear **cloudy or opaque**.

Lastly, medical solutions must be **physiologically compatible**, meaning they should not disturb the body's biochemical equilibrium or trigger harmful reactions. Factors like **pH, osmolarity**, and **electrolyte composition** must be carefully adjusted. Collectively, these properties make solutions safe and effective for a variety of therapeutic uses—from rehydration and drug administration to diagnostic applications.

Clinical Importance:

The clinical importance of solutions in healthcare cannot be overstated. They are foundational to numerous medical interventions, including **intravenous (IV) fluids**, which restore hydration and electrolyte balance; **oral medications**, where active ingredients are dissolved for optimal absorption; **irrigation solutions**, used during wound care and surgical procedures to reduce infection risk; and **parenteral nutrition**, which delivers essential nutrients directly into the bloodstream for patients unable to eat or absorb food orally. For nurses and allied health professionals, understanding solutions goes beyond passive knowledge—it is an active skill set. They must be able to **identify different types of solutions** (true solutions, colloids, suspensions), interpret and calculate **concentrations accurately**, and **prepare or administer solutions safely**, adhering to protocols that prevent contamination, incompatibility, or dosing errors. A strong command of solution chemistry directly supports safe patient care, reduces medication errors, and enables prompt, informed decision-making in dynamic clinical settings.

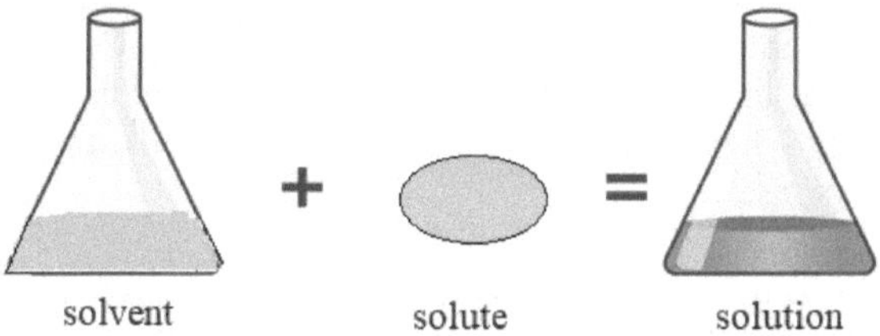

Figure 8.1 solution components

Types of Solutions

The clinical importance of solutions in healthcare cannot be overstated. They serve as the basis for a wide range of medical interventions, including **intravenous (IV) fluids**, which restore hydration and electrolyte balance; **oral medications**, in which active ingredients are dissolved to ensure effective absorption; **irrigation solutions**, used during wound care and surgical procedures to minimize infection risk; and **parenteral nutrition**, which delivers essential nutrients directly into the bloodstream for patients who are unable to eat or absorb nutrients through the gastrointestinal tract. For nurses and allied health professionals, competence in solution chemistry is not merely theoretical—it is a **practical and essential skill**. Healthcare providers must be able to **recognize different types of solutions** (true solutions, colloids, and suspensions), **understand and calculate concentrations** with precision, and **prepare or**

administer solutions safely, following established protocols to prevent contamination, chemical incompatibility, or dosing errors. Mastery of these concepts is critical to delivering safe, effective patient care, minimizing medical errors, and responding confidently in high-pressure clinical environments.

True Solutions

A true solution is a homogeneous mixture in which the solute is completely and uniformly dissolved in the solvent, forming a stable, single-phase system. The solute particles in true solutions are molecular in size—typically less than 1 nanometer—which makes them invisible to the naked eye and undetectable even under standard light microscopes. Due to their extremely small size and full dispersion, these solutions remain optically clear and transparent, do not settle out over time, and do not exhibit the Tyndall effect, which is the scattering of light by dispersed particles. One of the key distinguishing features of true solutions is their ability to pass freely through both filters and semipermeable membranes, unlike colloids and suspensions.

In clinical practice, true solutions are extensively used in intravenous therapy and medication administration. For instance, 0.9% sodium chloride, commonly known as normal saline, is frequently employed to restore hydration, dilute medications, and facilitate intravenous drug delivery. Another common example is 5% dextrose in water (D5W), which provides both fluids and a source of basic caloric energy. Ringer's lactate is a balanced electrolyte solution essential in fluid resuscitation and replacement therapy. Additionally, glucose-containing solutions are administered to treat hypoglycemia and to deliver energy in total parenteral nutrition regimens.

For nurses, understanding the nature and clinical applications of true solutions is critical to ensuring accurate preparation, proper labeling, and safe administration. Incompatible combinations may result in precipitation, chemical inactivation of drugs, or adverse patient reactions. Thus, checking for solution compatibility before mixing medications is an essential component of safe nursing practice.

Figure 8.2 types of solution

Colloids

Colloids represent a unique category of mixtures in which the dispersed particles are larger than those in true solutions but still small enough to remain suspended within the liquid medium without settling. These particles typically range from 1 to 1000 nanometers in diameter and, although they do not dissolve, they stay uniformly distributed throughout the solution. Colloidal mixtures are often cloudy or opalescent in appearance and are characterized by the Tyndall effect—the scattering of light by dispersed particles—a key feature that differentiates them from true solutions. Unlike solutes in true solutions, colloidal particles do not pass easily through semipermeable membranes and cannot be removed via ordinary filtration methods.

In clinical settings, colloids play a critical role in **fluid resuscitation and plasma volume expansion**, especially in cases of hypovolemia. Commonly used colloids include **albumin solutions** (5% or 25%), which are administered in shock, severe burns, or hypoalbuminemia. **Dextran**, a high-molecular-weight glucose polymer, is utilized in hemorrhagic shock to maintain vascular volume. **Synthetic colloids**, such as **hetastarch** and **gelatin-based products** like *Haemaccel* and *Gelofusine*, are often used as alternatives to crystalloids in critical care settings. **Lipid emulsions**, such as *Intralipid*, are employed in total parenteral nutrition (TPN) to supply essential fatty acids and caloric energy.

For nurses, the value of colloids lies in their ability to remain in the intravascular compartment longer than crystalloids, offering more prolonged plasma volume support. However, their use must be approached cautiously due to potential adverse effects, including allergic reactions, coagulation interference, and renal dysfunction—especially with synthetic variants. Nursing responsibilities include careful monitoring for **fluid overload, hypersensitivity reactions**, and ensuring compatibility with concurrently administered medications. Nurses must also document **infused volumes** and assess the patient's response, paying close attention to **blood pressure, urine output**, and **respiratory function** throughout the therapy.

Suspensions

Suspensions are heterogeneous mixtures where solid particles disperse in a liquid but remain undissolved. These particles are usually larger than 1000 nanometers, visible under a microscope, and tend to settle over time. Unlike true solutions and colloids, suspensions are unstable and require thorough shaking before administration for uniform dosing. They appear cloudy or opaque, do not pass through filters or semipermeable membranes, and are unsafe for intravenous use due to embolism risk.

Clinically, suspensions are used orally, topically, or rectally when high local drug concentration or delayed release is needed. Examples include antibiotic suspensions (amoxicillin-clavulanate), antacids (magnesium and aluminum hydroxide), barium sulfate for imaging, and activated charcoal for poisoning.

In nursing, suspensions must be shaken well before use to ensure dose accuracy, especially in children. Nurses should monitor side effects like constipation or diarrhea, check expiration dates after reconstitution, and educate patients or caregivers on proper use to prevent therapeutic failure or toxicity.

Table 8.1 Comparison of Solution Types in Clinical Practice

Feature	True Solution	Colloid	Suspension
Particle Size	< 1 nanometer	1–1000 nanometers	> 1000 nanometers
Appearance	Clear and transparent	Cloudy or milky	Opaque; particles may be visible
Stability	Very stable; does not settle	Relatively stable; does not settle quickly	Unstable; particles settle over time
Filtration	Passes through filters and membranes	May pass through filters; blocked by semipermeable membranes	Does not pass through filters
Light Scattering	Does not scatter light (no Tyndall effect)	Scatters light (shows Tyndall effect)	May scatter light, but usually not uniform
Clinical Route	Safe for intravenous (IV) use	Safe for IV (with caution, e.g., albumin)	Not used for IV; suitable for oral or topical use only
Examples in Healthcare	Normal saline, D5W, electrolyte solutions	Albumin, dextran, lipid emulsions	Antacid suspensions, antibiotic suspensions, barium sulfate
Nursing Consideration	Check compatibility and sterility	Monitor for allergic reactions or fluid overload	Shake well before administration; measure doses accurately

Concentrations

In clinical practice, accurately determining and adjusting the concentration of solutions is far from a routine task—it is a cornerstone of patient safety and therapeutic effectiveness. Whether preparing intravenous fluids, administering medications, or correcting electrolyte imbalances, understanding solution concentration enables healthcare providers to deliver the right substance in the appropriate form and dose. Even minor miscalculations can lead to treatment failure or serious harm, particularly in vulnerable populations such as neonates, elderly patients, or those in critical care.

Concentration refers to the quantitative relationship between the amount of solute—the substance being dissolved—and the volume of solvent or the entire solution. This ratio defines how "strong" or "dilute" a solution is, which in turn influences how the substance behaves in the body, including its absorption, distribution, metabolism, and excretion. Different clinical contexts demand different concentration levels, and choosing the correct one depends on both pharmacological requirements and the patient's physiological status.

Healthcare professionals rely on several units to express concentration, tailored to specific clinical tasks. Some are based on **mass**, such as grams per 100 milliliters (g/100 mL); others use **molarity** (mol/L) to quantify the number of moles of solute per liter of solution; and in electrolyte management, **milliequivalents per liter (mEq/L)** are used to reflect the chemical reactivity or charge. Mastery of these units—and the ability to convert between them—is essential for safe drug preparation, calculating IV flow rates, preparing rehydration fluids, and accurately interpreting lab values.

This section introduces the most commonly used concentration units in clinical chemistry and illustrates their practical applications in real healthcare scenarios.

Types of Concentration

In healthcare and pharmaceutical settings, the **type of concentration unit** used depends on the nature of the substance being administered, the required precision, and the clinical context. Below are the most common types of concentration used in clinical practice:

Mass/Volume Percent (% w/v)

Mass/Volume Percent (% w/v) is one of the most commonly used concentration units in healthcare. It expresses the mass of solute (in grams) per 100 milliliters of solution.

Formula: Mass/Volume Percent (% w/v)

$$\%\,w/v = \frac{\text{grams of solute}}{\text{milliliters of solution}} \times 100$$

Where:

- **% w/v** = concentration in percent
- **grams of solute** = the amount of solid substance dissolved
- **milliliters of solution** = the total final volume of the liquid solution.

Example: To prepare **100 mL of a 5% (w/v) glucose solution**:

$$\textbf{grams of glucose} = \frac{\mathbf{5}}{\mathbf{100}} \times \mathbf{100} = \mathbf{5\ grams}$$

So, dissolve **5 grams** of glucose in enough water to make the final volume **100 mL.**

Clinical Examples of Common Solutions

Several solutions used in healthcare follow standard concentration formats that nurses must understand for accurate preparation and administration. For instance, **5% dextrose in water (D5W)** contains **5 grams of dextrose per 100 milliliters** of water, which equates to **50 grams of dextrose per 1 liter** of solution. It is commonly used for fluid replacement and to provide minimal caloric support. Similarly, **0.9% sodium chloride**, also known as **normal saline**, consists of **0.9 grams of NaCl per 100 milliliters** of water, translating to **9 grams per liter**.

This isotonic solution is widely used for intravenous hydration, electrolyte balance, and medication dilution. Another common formulation is **1% lidocaine injection**, which contains **1 gram of lidocaine in 100 milliliters** of solution. It is used for local anesthesia and sometimes in cardiac emergencies. Understanding these percentage concentrations is essential for ensuring accurate dosing, safe dilution practices, and proper documentation in clinical settings.

Nursing Relevance of % w/v Concentrations

In everyday nursing practice, the **percent weight/volume (% w/v)** unit is one of the most frequently encountered concentration formats, especially on intravenous fluid bags, topical medications, and liquid oral drugs. This unit expresses the number of grams of solute in every 100 milliliters of solution. A clear understanding of % w/v is crucial for **safe, accurate, and accountable nursing care**.

Proficiency in interpreting and calculating % w/v concentrations allows nurses to verify medication dosages, follow pharmacy instructions precisely, and ensure proper dilution before administration. For instance, knowing that a 5% dextrose solution contains 5 grams of glucose in every 100 mL is essential for tracking both fluid and caloric intake—especially in vulnerable populations.

A frequent error involves assuming that simply adding 100 mL of solvent to a known amount of solute yields a 100 mL solution. In fact, the final volume must be adjusted after dissolution to exactly 100 mL to ensure accurate concentration. Misjudgments in this area can lead to **underdosing or overdosing**, which may have serious consequences, particularly in **pediatric patients, critical care settings, and intravenous drug delivery**. As frontline providers, nurses serve as the critical link between prescription and patient, making mastery of % w/v concentrations not just a mathematical task—but a clinical responsibility.

Example Problem:

A nurse is preparing an antiseptic solution for wound irrigation. The order requires **250 mL of a 0.5% (w/v) iodine solution**. The iodine comes in powder form.

Question:
How many grams of iodine powder are needed to prepare 250 mL of a 0.5% (w/v) solution?

Solution :

$$\text{grams of solute} = \frac{0.5 \times 250}{100} = 1.25 \text{ grams}$$

Answer: The nurse needs to dissolve **1.25 grams of iodine** in water to make **250 mL** of a 0.5% solution.

Check Yourself
How many grams of glucose are needed to prepare **500 mL** of a **10% (w/v)** glucose solution?

Volume/Volume Percent (% v/v)

Volume/Volume Percent (% v/v) expresses the **volume of solute (in mL)** per **100 mL of solution**. It is typically used when both the solute and solvent are **liquids**, such as in alcohol-based preparations or liquid medications.

Definition:

$$\% \, v/v = \frac{ml \ of \ solute}{ml \ of \ solution} \times 100$$

Clinical Examples of % v/v Solutions

In healthcare settings, **percent volume/volume (% v/v)** is used to express the concentration of liquid solutes dissolved in liquid solvents, a format commonly applied to **alcohol-based solutions and liquid medications**. For example, **70% ethanol** refers to a solution containing **70 milliliters of ethanol per 100 milliliters of final solution**. This concentration is widely used as a **disinfectant and antiseptic**, offering optimal antimicrobial activity in clinical and surgical environments.

Other examples include **tinctures and oral liquid medications**, which are often formulated using % v/v to ensure accurate dosing and enhance solubility of active ingredients. Understanding % v/v is essential for nurses when administering or preparing such solutions, particularly in wound care, topical therapies, and pediatric dosing, where precision and compatibility are critical. Familiarity with this unit also supports safe communication with pharmacists and ensures compliance with medical protocols regarding fluid volume and concentration.

Nursing Relevance of % v/v Concentrations

Understanding **percent volume/volume (% v/v)** concentrations is essential for nurses involved in medication administration, topical treatments, and solution preparation. This unit is commonly used in **topical antiseptics**, such as ethanol or isopropyl alcohol, as well as in **inhalation therapies** and **liquid compounding in pharmacy settings**. Nurses must be especially vigilant when working with **concentrated alcohol-based products**, as improper dilution can lead to **tissue irritation, toxicity, or reduced therapeutic efficacy**.

For example, some formulations of **chlorhexidine or ethanol** must be diluted to precise strengths before use on mucous membranes or open wounds. Therefore, it is critical to **carefully check the labeled concentration and follow dilution instructions** exactly as indicated. Failure to do so may compromise patient safety. Mastery of % v/v concepts ensures accurate preparation, supports safe clinical decision-making, and minimizes the risk of adverse reactions during routine and emergency care.

Example Problem:

Question:
How much **95% ethanol** is needed to prepare **250 mL** of a **70% (v/v)** ethanol solution?

This is a dilution problem. You can use the dilution formula:

$$C_1V_1=C_2V_2$$

Where:

- C1 = original concentration (95%)
- V1 = volume of stock to use
- C2 = desired concentration (70%)
- V2 = final volume (250 mL)

Solution:

95% × V1 = 70% × 250

$$\boldsymbol{V1 = \frac{70 \times 250}{95} = 184.2\ ml}$$

Answer: Use **184.2 mL of 95% ethanol**, and add water to make a final volume of **250 mL.**

Check Yourself

You are asked to prepare **150 mL of a 40% (v/v)** alcohol solution using 70% stock.
How many milliliters of the 70% stock do you need?

Molarity (M)

Molarity is a fundamental unit of concentration in chemistry and clinical laboratories. It expresses the number of **moles of solute** dissolved in **one liter of solution.**

Definition:

$$\boldsymbol{Molarity\ (M) = \frac{moles\ of\ solute}{liters\ of\ solution}}$$

Mole = the amount of a substance that contains 6.022×10^{23} particles (atoms, ions, or molecules).

Used in: laboratory solutions, IV electrolyte preparations, buffer systems, and medication compounding.

Clinical Example: Preparing a Molar Solution

In a clinical or laboratory setting, preparing a **1 molar (1 M) solution** of sodium chloride (NaCl) requires dissolving **1 mole of NaCl**—equivalent to approximately **58.5 grams**—in water, and then adjusting the total volume to exactly **1 liter**. It is important to first dissolve the solute in a partial volume of water, then add additional solvent to bring the solution up to the final volume. Adding solute directly to 1 liter of water would increase the total volume beyond 1 liter, resulting in a concentration lower than intended.

Understanding how to correctly prepare molar solutions is essential for accurate **reagent formulation**, adjusting **electrolyte concentrations**, and interpreting **laboratory prescriptions** in both clinical chemistry and pharmacology.

Nursing Relevance of Molarity

For nurses, **molarity** plays a critical role in understanding the concentration of substances used in **diagnostic, therapeutic, and pharmaceutical settings**. It is commonly encountered in laboratory reports and during the preparation of specialized IV solutions. Molarity is especially important when calculating **electrolyte adjustments**, creating **buffer systems**, or preparing **pH-specific solutions**.

In dialysis, critical care, and parenteral nutrition, precise molar calculations are vital to ensure that **electrolyte and acid–base balance** is maintained. A strong grasp of molarity allows nurses to interpret lab data more effectively, make safer clinical decisions, and collaborate competently with pharmacists and physicians in multidisciplinary teams. Ultimately, mastering this concept enhances the accuracy and safety of patient care in chemistry-driven clinical environments.

Example Problem:

Question:
How many grams of **KCl** (molar mass = 74.5 g/mol) are needed to prepare **500 mL of a 0.2 M solution**?

Step 1: Use the molarity formula:

$$Molarity\ (M) = \frac{moles\ of\ solute}{liters\ of\ solution}$$

Then: moles = M × V =0.2 × 0.5 = 0.1 mol

Step 2: Convert moles to grams:

Grams = moles × molar mass = 0.1 × 74.5 =7.45 g

Answer: Dissolve **7.45 g of KCl** in enough water to make **500 mL** of solution.

Check Yourself

How many grams of NaOH (40 g/mol) are needed to prepare 250 mL of a 0.1 M solution?

Milliequivalents per Liter (mEq/L)

Milliequivalents per liter (mEq/L) is a unit of concentration commonly used to express the amount of **electrically charged ions (electrolytes)** in a liter of solution—especially in blood plasma and intravenous fluids.

This unit reflects both the **amount of substance** and its **electrical charge (valence)**, making it particularly important in **fluid and electrolyte balance, acid–base regulation**, and **IV therapy**.

Formula:

$$mEq = (\frac{mg\ of\ substance\ \times valence}{molar\ mass})$$

mEq/L= Milliequivalents per liter

Table 8.2 Key Electrolyte Valences:

Electrolyte	Symbol	Valence
Sodium	Na^+	1
Potassium	K^+	1
Calcium	Ca^{2+}	2
Magnesium	Mg^{2+}	2
Chloride	Cl^-	1
Bicarbonate	HCO_3^-	1

Clinical Example: Blood Sodium Concentration in mEq/L

In clinical practice, electrolyte concentrations are commonly expressed in milliequivalents per liter (mEq/L) to reflect both the amount of substance and its chemical activity. For example, the normal blood sodium level ranges from 135 to 145 mEq/L, meaning that each liter of plasma contains 135 to 145 milliequivalents of sodium ions (Na^+). This unit accounts for both the number of particles and the valence (charge) of the ion, making it particularly useful for evaluating ionic balance in body fluids.

Monitoring sodium levels is critical in managing conditions such as dehydration, fluid overload, hyponatremia, and hypernatremia. Understanding mEq/L allows nurses to interpret lab results, adjust IV electrolyte therapy appropriately, and respond to electrolyte imbalances with informed clinical judgment.

Why mEq/L Matters in Healthcare

The use of **milliequivalents per liter (mEq/L)** is essential in healthcare for accurately monitoring and correcting **electrolyte imbalances**, particularly in patients with conditions such as **dehydration, renal impairment, acid-base disorders, or cardiac dysfunction**. This unit reflects not only the quantity of an ion but also its **electrical charge**, making it ideal for clinical applications. For example, **0.9% sodium chloride (normal saline)** contains **154**

mEq/L of sodium (Na^+) and **154 mEq/L of chloride (Cl^-)**, providing a balanced isotonic solution used routinely in IV therapy.

Nurses must also apply this concept when calculating and administering **electrolyte replacements**, such as potassium, calcium, and magnesium, where dosing precision is critical to avoid serious complications like **arrhythmias, neuromuscular disturbances, or metabolic instability**. Proficiency in interpreting and using mEq/L values enhances the nurse's ability to make **evidence-based decisions** and deliver safe, effective patient care in both acute and chronic settings.

Example Problem:

Question:
How many mEq are in **1 gram of KCl** (molar mass = 74.5 g/mol, K^+ has valence 1)?

$$\text{mEq} = \frac{1000 \text{ mg } \times 1}{74.5} = 13.42 \text{ m Eq}$$

Answer: 1 gram of KCl contains **13.42 mEq** of potassium.

Check Yourself

How many mEq of **Ca^{2+}** are in **400 mg** of calcium (molar mass = 40 g/mol, valence = 2)?

Normality (N)

Normality is a unit of concentration used to express the **reactive capacity** of a solution. It is especially useful in acid–base chemistry and redox reactions, where the **chemical reactivity (equivalents)** of a solute matters more than just its molar amount.

Definition:

$$\boldsymbol{Normality\ (N) = \frac{equivalents\ of\ solute}{liters\ of\ solution}}$$

Normality vs. Molarity

In clinical and laboratory chemistry, both **molarity (M)** and **normality (N)** are used to express solution concentration, but they serve different purposes. **Molarity** refers to the number of **moles of solute per liter of solution** and is a basic unit of concentration. **Normality**, on the other hand, accounts for the **reactive capacity** of a solute by incorporating its **valence**—the number of equivalents that a substance can donate or accept in a chemical reaction.

For example, **sulfuric acid (H_2SO_4)** dissociates to release **two hydrogen ions (H^+)** per molecule, so a **1 M solution of H_2SO_4** is **2** N, because each mole provides two equivalents of H^+. In contrast, **sodium hydroxide (NaOH)** provides **one hydroxide ion (OH^-)** per molecule, so a **1 M NaOH solution** is also **1 N**.

Normality is especially useful in **acid–base titrations, buffer preparation**, and **electrolyte calculations**, where the number of reactive ions directly influences the outcome of the reaction or treatment.

Molarity (M) measures the number of moles of solute per liter.

Normality (N) adjusts for **valence** (the number of reactive units per molecule).

$$\textbf{Normality} = \textbf{Molarity} \times \textbf{n-factor}$$

Examples: Molarity vs. Normality

- For **H_2SO_4 (sulfuric acid)**, which donates **2 H^+ ions per molecule**, the normality is **twice the molarity**:
  - 1 M H_2SO_4 = 2 N (because each mole provides 2 equivalents of H^+)
- For **NaOH (sodium hydroxide)**, which donates **1 OH^- ion per molecule**, the normality is **equal to the molarity**:
  - 1 M NaOH = 1 N (each mole provides 1 equivalent of OH^-)

Clinical Relevance of Normality

While not commonly labeled on IV bags, **normality (N)** plays a critical role in various **clinical and laboratory applications**. It is frequently used in **acid–base titrations** to assess the concentration and reactivity of substances such as **gastric juices, blood buffers, and pharmaceutical formulations**. In hospital and research settings, normality is also essential for **calculating the concentration of buffers** used in IV solutions, diagnostic assays, and pH-sensitive drug preparations.

Understanding normality enables healthcare professionals to interpret lab results involving **reactive ions**, particularly in situations where **ionic strength or the number of active ions** directly influences biological function or therapeutic effectiveness. Although less visible in bedside practice, normality remains a fundamental concept in **clinical chemistry, pharmacy compounding, and analytical diagnostics**, supporting the scientific basis for safe and effective care.

Example Problem:

Question:
What is the normality of a 1.5 M H_2SO_4 solution?

- H_2SO_4 donates **2 protons (H^+)** → n-factor = 2

$$\boldsymbol{Normality\ (N) = 1.5 \times 2 = 3\ N}$$

Check Yourself

What is the normality of 0.25 M $Ca(OH)_2$?

Nursing Tip

Nurses should be comfortable using **mEq/L** when interpreting **electrolyte panels**, as this unit reflects both the quantity and electrical charge of ions in body fluids—essential for managing conditions like dehydration, acidosis, and electrolyte imbalance. In addition, understanding **molarity and normality** becomes especially important when working in **clinical laboratories**, assisting with **buffer preparation**, or supporting **acid–base titration procedures**. This foundational knowledge helps ensure accuracy in medication compounding, diagnostic testing, and critical care interventions.

Table 8.3 Comparison of Molarity, Normality, and mEq/L

Feature	Molarity (M)	Normality (N)	Milliequivalents per Liter (mEq/L)
Definition	Moles of solute per liter of solution	Equivalents of solute per liter of solution	Milligrams of ion × valence ÷ molecular weight
Focus	Amount of substance	Reactive units (acid/base or redox capacity)	Electrical activity of charged particles (ions)
Used For	General chemistry, lab solutions	Acid–base titrations, redox reactions	Electrolyte concentrations in blood or IV fluids
Considers valence?	No	Yes	Yes
Unit Symbol	M (mol/L)	N (eq/L)	mEq/L
Example Substance	1 M HCl = 1 mol HCl/L	1 M H_2SO_4 = 2 N (since it donates 2 H^+ ions)	1 g KCl ≈ 13.4 mEq (K^+ has valence = 1)
Common in Clinical Settings?	Yes (labs, IV prep)	□ Rarely (mostly in labs)	Very common (labs, IVs, cardiac, renal care)

Parts Per Million (ppm)

Parts per million (ppm) is a unit of concentration used to measure very **small quantities of a substance** in a large amount of solvent—typically water, air, or blood. It is especially useful when dealing with **trace elements**, **toxins**, **pollutants**, or **nutrients** found at low levels.

Definition:

$$PPM = \frac{mg\ of\ solute}{liters\ of\ solution}$$

In water-based solutions:

- 1 ppm = 1 mg of solute per liter of solution
- 1 ppm = 1 part per 1,000,000 parts of solution

Clinical and Public Health Examples of ppm and ppb

In both clinical and public health settings, the concentrations of trace substances are often measured using **parts per million (ppm)** and **parts per billion (ppb)**. These units are essential when monitoring extremely low levels of substances that, despite their small quantities, can have significant physiological or environmental impacts.

For instance, **fluoride** levels in municipal drinking water are maintained between **0.7 and 1.2 ppm** to strengthen tooth enamel and reduce the prevalence of dental cavities. On the other hand, **carbon monoxide (CO)**, a dangerous gas, becomes hazardous at concentrations exceeding **35 ppm**, prompting the use of safety alarms in hospitals, ambulances, and enclosed public spaces.

In toxicology, trace amounts of **heavy metals** such as **lead and mercury** are reported in either ppm or ppb when measured in **blood, urine, or water samples**, as even minimal exposure can result in neurological damage, particularly in children and pregnant women. Similarly, in **infection control**, **chlorine** is used to disinfect hospital water systems at concentrations ranging between **0.2 and 1.0 ppm**, helping to prevent microbial contamination and protect vulnerable patients.

For nurses and allied health professionals, understanding ppm and ppb is vital when working in **public health, environmental monitoring, occupational safety, infection control, or toxicology**. These small units represent critical thresholds—minor deviations can indicate emerging danger or the need for urgent intervention.

Relevance of ppm and ppb in Healthcare

Measurements in **parts per million (ppm)** and **parts per billion (ppb)** are indispensable across several healthcare domains, including **toxicology, environmental health**, and **infection control**. These ultra-low concentration units are used to evaluate the **safety, quality, and purity** of critical substances such as air, water, medications, and biological fluids.

For example, ppm-based monitoring is vital when assessing **lead concentrations in children**, **chlorine levels** in hospital water systems, or **carbon monoxide exposure** in respiratory care and emergency services. Even slight deviations from safety thresholds can indicate potential toxicity or contamination, especially in vulnerable populations.

Nurses play a central role in applying these measurements—**interpreting laboratory results**, **educating patients and caregivers** about exposure risks, and **advocating for public health policies** that uphold environmental safety. Whether in hospitals, public health units, or community settings, understanding and applying ppm and ppb concepts enables nurses to support **early detection, preventive care**, and **evidence-based clinical decisions**. Ultimately, these competencies help protect both **individual health** and **population-level outcomes**.

Example Problem:

Question:
A water sample contains **2 mg of lead** in **1 liter**. What is the concentration in ppm?

$$\boldsymbol{PPM} = \frac{\boldsymbol{2\ mg\ of\ lead}}{\boldsymbol{1\ liters\ of\ solution}} = 2\ ppm$$

Check Yourself: A toxicology report shows that a patient's blood contains 3 mg of a pesticide in 2 liters of blood. What is the concentration in ppm?

Preparation Techniques for Clinical Solutions

The **accurate and sterile preparation** of clinical solutions is a fundamental responsibility in nursing and healthcare practice. Whether formulating an **intravenous (IV) fluid, oral medication**, or **wound irrigation solution**, strict adherence to preparation protocols is critical to ensure **correct solute concentration, aseptic conditions**, and **compliance with regulatory standards**. Each step—from selecting the correct solute and solvent to measuring, mixing, labeling, and storing the solution—must be performed with precision. Even minor errors in dilution or contamination can result in **underdosing, drug toxicity, or severe infections**, especially in vulnerable populations such as neonates, immunocompromised patients, or the critically ill. Mastery of solution preparation techniques protects patient safety, maintains therapeutic efficacy, and upholds the professional standards required in all clinical settings.

Preparation of 0.9% Sodium Chloride Solution (Normal Saline)

Normal saline (0.9% NaCl) is one of the most widely used solutions in clinical practice. It is commonly employed for **intravenous hydration, wound irrigation**, and **dilution of injectable medications**. As an **isotonic solution**, it contains **0.9 grams of sodium chloride per 100 milliliters of sterile water**, which closely mimics the osmolarity of blood plasma, making it safe for a broad range of medical applications.

To prepare **100 mL of 0.9% NaCl solution**, the following steps should be followed precisely:

1. **Accurately weigh 0.9 grams of sodium chloride (NaCl)** using a calibrated analytical balance.
2. **Dissolve the NaCl in approximately 80 mL of sterile water for injection**, gently stirring to ensure complete dissolution.
3. Once fully dissolved, **add additional sterile water to bring the final volume up to exactly 100 mL**. It is important to adjust to final volume, not just add 100 mL of water initially, as this would result in incorrect concentration.
4. **Maintain strict aseptic technique** throughout the procedure to prevent microbial contamination.
5. **Label the container clearly**, including:

- Name of the solution (0.9% Sodium Chloride)
- Concentration
- Date and time of preparation
- Initials or full name of the person preparing the solution

Errors in preparation—such as incorrect concentration, contamination, or improper labeling—can lead to **serious clinical consequences**, including electrolyte imbalances or infections. Therefore, precise technique, proper documentation, and adherence to institutional protocols are essential when preparing sterile solutions for patient use.

Preparation of 5% Dextrose in Water (D5W)

Dextrose solutions, such as **5% dextrose in water (D5W)**, are commonly used in clinical practice for **fluid replacement**, **providing a source of calories**, and as a **vehicle for drug infusion**. A 5% solution means that there are **5 grams of dextrose per 100 milliliters**, or **50 grams per liter** of final solution.

To prepare **1 liter of 5% dextrose solution**, follow these steps:

1. **Weigh exactly 50 grams of glucose (dextrose monohydrate)** using a calibrated balance.
2. **Transfer the dextrose into a clean mixing container**, and add **approximately 800 to 900 milliliters of sterile water for injection**.
3. **Stir or shake gently** to ensure the dextrose is fully dissolved. Complete dissolution is necessary to ensure accurate dosing and prevent crystallization.
4. Once fully dissolved, **add more sterile water to bring the total volume up to exactly 1000 milliliters (1 liter)**. Never assume that adding 1 liter of water to 50 g of solute will yield a 1-liter solution.
5. **Label the solution clearly**, indicating:

    - Solution name (5% Dextrose in Water)
    - Concentration
    - Date and time of preparation
    - Preparer's initials

If the solution is **not prepared in a sterile compounding facility**, it should be used **promptly** and discarded if sterility cannot be maintained. Dextrose solutions can serve as a **nutrient source** in patients who are nil per os (NPO) or undergoing supportive therapy but must be handled with **strict aseptic precautions** to avoid infection, especially in intravenous administration.

Table 8. 4 Common Glucose Solutions:

Solution Name	Concentration	Use
D5W	5% glucose	IV hydration with mild energy supply
D10W	10% glucose	Higher caloric needs or neonatal use
D50W	50% glucose	Emergency treatment of hypoglycemia

Preparing a Hypotonic Saline Solution

A common clinical task for nurses involves the preparation and administration of **intravenous (IV) fluids** tailored to a patient's specific needs. For example, when a physician prescribes **1,000 mL of 0.45% sodium chloride (NaCl)**—a **hypotonic saline solution**— to a dehydrated patient requiring gentle cellular rehydration, the nurse must apply core concepts of solution chemistry to ensure accuracy and safety. A 0.45% NaCl solution contains **0.45 grams per 100 milliliters**, equivalent to **4.5 grams per liter**. To prepare this solution from sodium chloride powder, the nurse accurately **weighs 4.5 grams of NaCl**, dissolves it in **approximately 800–900 mL of sterile water for injection**, and then adjusts the final volume to exactly **1,000 mL**. The solution must be **clearly labeled** with the name, concentration, total volume, date, and preparer's initials. If not compounded under sterile pharmacy conditions, it should be used promptly to minimize contamination risk.

Beyond hydration, solutions such as this are used in a wide range of **clinical applications**, including **drug dilution, electrolyte replacement, dose adjustments**, and **parenteral nutrition**. Nurses must also assess the **compatibility** of solutions when co-administered with medications, as mixing incompatible agents may lead to **precipitation, reduced drug efficacy**, or **serious adverse**

reactions. This example illustrates how an understanding of **solution concentration, sterility, and chemical compatibility** is essential for ensuring **safe, evidence-based nursing practice** in both acute and routine care settings.

Sterility and Safety Guidelines in Solution Preparation

When preparing solutions intended for **parenteral administration**—including intravenous (IV), intramuscular (IM), and subcutaneous (SC) routes—**strict aseptic technique** is essential to prevent microbial contamination and safeguard patient health. Maintaining sterility is a non-negotiable standard, as contaminated solutions may lead to **serious infections**, **sepsis**, or **life-threatening drug reactions**.

All components involved in preparation must be sterile, including **water for injection, syringes, mixing equipment**, and containers. Workstations must be thoroughly **cleaned and disinfected** before and after each preparation session. Personnel are required to wear **appropriate personal protective equipment (PPE)**—such as sterile gloves, face masks, and gowns—particularly when working in clean areas or handling open vials and ampules.

To maintain aseptic conditions, **direct contact** with the interior surfaces of caps, syringes, vial stoppers, or IV bags must be strictly avoided. In institutional settings like hospitals or infusion centers, solution preparation should ideally be performed by **trained pharmacy staff** in a **certified laminar airflow hood** or **cleanroom environment**, where air filtration and controlled airflow reduce the risk of contamination.

For nurses, understanding and adhering to these **sterility protocols** is critical, especially when managing **high-risk medications**, **IV admixtures**, or **compounded sterile products (CSPs)**. Observing these guidelines ensures not only regulatory compliance but also protects patients from avoidable harm and enhances the overall safety of clinical care.

Safety Tips for Healthcare Providers

The safe preparation and administration of clinical solutions demand strict attention to detail and unwavering adherence to established best practices. Healthcare providers must **double-check all calculations**, including solute amounts, concentrations, and dilution ratios, to minimize the risk of dosing errors. Equally important is verifying the correct **type of solvent** and ensuring that the **final volume** is accurately achieved—not approximated—to maintain intended solution strength.

Each prepared solution must be **clearly and completely labeled** with essential information: the **name of the substance**, its **concentration**, **total volume**, **date and time of preparation**, and the **initials or signature** of the person who prepared it. Unlabeled or improperly labeled solutions pose a serious threat to patient safety and must **never be administered** under any circumstances.

Before use, solutions should undergo **visual inspection**. Any solution that appears **cloudy, discolored**, contains **visible particles or precipitates**, or is **past its expiration date** must be discarded immediately, as these are indicators of **contamination or chemical degradation**.

By consistently following these safety practices, healthcare professionals uphold the **highest standards of care**, reduce the risk of harm, and contribute to a safer clinical environment for patients and staff alike.

Clinical Applications of Solutions

In healthcare, **solutions serve far more than hydration purposes**—they are vital tools in the delivery of comprehensive patient care. Solutions are used to **administer medications intravenously or orally**, **correct electrolyte imbalances**, **support nutritional needs** through parenteral nutrition, and **manage both acute and chronic medical conditions**, including shock, renal failure, and infections. Intravenous fluids such as **normal saline, dextrose solutions, and balanced electrolyte formulations** enable precise control over fluid volume and composition in critically ill patients.

In addition, medications are often dissolved or diluted in specific solutions to facilitate **safe infusion**, achieve **targeted pharmacological effects**, or prevent **tissue irritation**. Proper understanding of **solution concentration, stability, sterility, and compatibility** is essential to prevent adverse events such as **precipitation, inactivation, or toxicity**. For nurses and healthcare providers, mastery of solution preparation and administration is central to ensuring **safe, effective, and evidence-based clinical care**.

IV Fluids (Intravenous Solutions)

Intravenous (IV) fluids are essential components of modern clinical care, used to maintain hydration, correct electrolyte imbalances, deliver medications, and support circulation. These fluids are broadly **categorized based on their tonicity**—that is, their osmotic relationship to blood plasma—and their **chemical composition**. The three main types of IV fluids include **isotonic solutions** (e.g., 0.9% NaCl, Lactated Ringer's), which have a similar osmolarity to blood and are used for volume expansion without causing fluid shifts; **hypotonic solutions** (e.g., 0.45% NaCl), which have lower osmolarity and are used to rehydrate cells in conditions like hypernatremia; and **hypertonic solutions** (e.g., 3% NaCl, D5NS), which have higher osmolarity and draw fluid out of cells, often used in critical care to manage cerebral edema or hyponatremia.

Understanding these classifications helps nurses and healthcare providers select the appropriate fluid therapy for each clinical scenario, ensuring effective and safe patient management.

Table 8.5 fluids category based on tonicity and composition

Type	Example	Clinical Use
Isotonic	0.9% NaCl, D5W, LR	Hydration, fluid resuscitation
Hypotonic	0.45% NaCl	Cellular dehydration, hypernatremia
Hypertonic	D10W, D5NS, 3% NaCl	Severe hyponatremia, hypoglycemia emergencies

Nursing Tip: Before initiating or adjusting IV fluid therapy, always assess the patient's vital signs, renal function, and electrolyte status. These parameters are critical for selecting the appropriate fluid type, preventing complications such as fluid overload or electrolyte imbalance, and ensuring safe, individualized patient care.

Drug Solutions and Dosage Calculations

In clinical practice, many medications are administered in **solution form**, requiring nurses and healthcare providers to perform **precise dosage calculations** to ensure safe and effective treatment. This is especially important when **diluting concentrated drugs**, such as antibiotics or electrolytes, where small errors can lead to toxicity or underdosing. Additionally, **weight-based dosing**—commonly used in **pediatric care**, **chemotherapy**, and **critical care**—requires accurate calculations based on the patient's body weight (e.g., mg/kg).

In continuous or intermittent infusion therapies, adjusting the **infusion rate** (e.g., in mL/hour or mg/min) is essential for maintaining therapeutic drug levels while preventing complications such as hypotension, arrhythmias, or organ damage. Mastery of these calculations supports clinical decision-making and upholds patient safety in both acute and long-term care settings.

Basic formula for dosage calculation:

Dose (mg) =Volume (mL)×Concentration (mg/mL)

Example: A patient needs 500 mg of a drug. You have a solution labeled 250 mg/5 mL.
How many mL do you give?

Solution : $\boldsymbol{ml} = \frac{\mathbf{500}\,\boldsymbol{mg}}{\mathbf{250}\,\boldsymbol{mg}} \times 5\ ml\ = 10\ ml$

Compatibility and Mixing Considerations

In clinical settings, **mixing incompatible solutions or medications** can lead to serious complications, including **precipitation, loss of drug activity, or toxic reactions**. Understanding and applying

compatibility principles is essential for preventing adverse outcomes. Healthcare providers must always **consult up-to-date IV compatibility charts** before combining medications with IV fluids or other drugs in the same line. For example, **calcium should not be mixed with phosphate**, as this combination poses a high risk of **crystal formation and embolism**. Additionally, **heparin** is known to be **incompatible with many antibiotics**, potentially reducing their efficacy or causing line occlusion.

For **high-risk medications** such as **total parenteral nutrition (TPN)**, **vasopressors**, and **chemotherapy agents**, the use of **dedicated IV lines** is strongly recommended to prevent cross-contamination and ensure optimal drug stability. Adhering to these compatibility guidelines is critical for **maintaining medication safety, protecting vascular access, and ensuring therapeutic effectiveness**.

Nursing Practice in Solution Safety

Safe preparation and administration of solutions require vigilant **nursing practice** grounded in accuracy and observation. Nurses must **always read drug labels carefully** to confirm the correct medication, concentration, and solvent before mixing or administering any solution. It is equally important to **visually inspect all solutions** for signs of instability, such as **cloudiness, crystallization, or unexpected color changes**, which may indicate contamination or incompatibility.

Every prepared solution should be **clearly labeled** with the **drug name, dose, diluent or solvent used**, and the **date and time of preparation**, along with the initials of the preparer. This not only supports safe clinical workflow but also ensures traceability and compliance with institutional and legal standards. Adhering to these practices reinforces a culture of safety and accountability in medication management.

Clinical Scenario Example: Safe Administration of IV Potassium Chloride

You are preparing to administer **intravenous potassium chloride (KCl)** as ordered by the physician: **20 mEq diluted in 100 mL of 0.9% sodium chloride (normal saline)** to be infused over **1 hour**. This scenario highlights several critical nursing considerations. First and foremost, **potassium chloride must never be administered undiluted or given as an IV push**, as this can lead to life-threatening cardiac arrhythmias. The solution must be thoroughly mixed and infused slowly using an infusion pump to ensure precise delivery. **Continuous cardiac monitoring is essential**, especially in patients with underlying heart conditions, as changes in serum potassium can significantly affect the heart's electrical activity.

Renal function must be assessed before and during potassium administration, as impaired kidneys may reduce potassium excretion and increase the risk of hyperkalemia. Nurses should also verify IV compatibility with other medications or fluids to prevent harmful interactions. This highlights the importance of clinical judgment, chemistry knowledge, and adherence to safety protocols when administering high-alert drugs like potassium.

Table 8.6 : **Summary of** Common IV Fluids and Their Clinical Applications

Fluid	Type	Components	Tonicity	Clinical Uses
0.9% NaCl	Crystalloid	Na^+, Cl^-	Isotonic	Resuscitation, shock, hyponatremia
D5W	Crystalloid	Dextrose in water	Isotonic → Hypo*	Glucose supply, hypernatremia dehydration
LR	Crystalloid	Na^+, Cl^-, K^+, Ca^{2+}, lactate	Isotonic	Surgery, burns, GI losses
0.45% NaCl	Crystalloid	Half-strength saline	Hypotonic	Cellular dehydration, hypernatremia
D5NS	Crystalloid	Dextrose + 0.9% NaCl	Hypertonic	Maintenance, calories + electrolytes
D10W	Crystalloid	Dextrose	Hypertonic	Hypoglycemia, TPN base
3% NaCl	Crystalloid	High-concentration saline	Hypertonic	Severe hyponatremia, cerebral edema
Colloids e.g., albumin	Colloid	Proteins/polysaccharides	Varies	Volume expansion, hypoalbuminemia

***Notes**:1. D5W is isotonic in the bag but becomes hypotonic once glucose is metabolized 2. Always monitor electrolytes, renal function, and fluid status when administering IV fluids.*

Chemistry Link to Health

Enhancing Patient Safety: The Role of Chemistry and Smart Infusion Systems

A June 2024 review in the *Journal of Patient Safety* highlights the shift toward systems-based risk management to improve IV medication safety. Advanced tools like smart infusion pumps and automated preparation systems reduce human errors, standardize drug concentrations, and enable real-time monitoring.

These technologies control infusion rates precisely and alert clinicians to drug incompatibilities, while automated compounding ensures accurate sterile preparation of complex solutions like TPN. However, they do not replace the essential nursing knowledge of chemistry and pharmacology—understanding concentration, solubility, compatibility, and stability remains vital.

Effective use of these systems requires combining technical skills with clinical judgment; even smart pumps cannot prevent errors if incompatible drugs are mixed beforehand. Thus, integrating chemical literacy with sterile technique and technology proficiency empowers nurses to reduce errors and improve patient outcomes.

Clinical chemistry is therefore a crucial part of daily nursing practice to safely and effectively use advanced IV technologies.

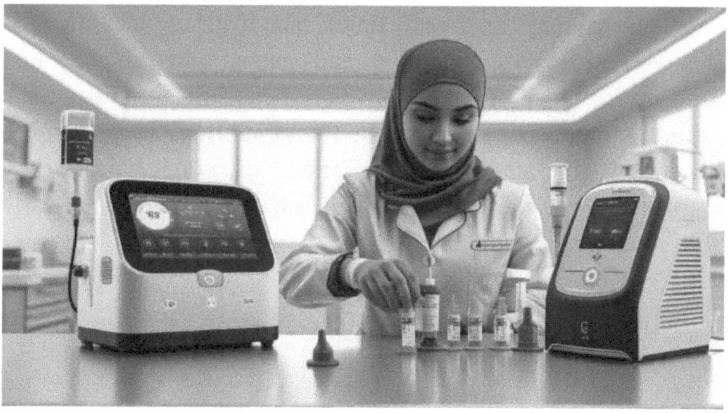

Chapter Summary

This chapter highlights the essential role of solutions in clinical care and equips nursing students with the foundational knowledge and practical skills to prepare and use them safely. It begins by explaining the main types of solutions—true solutions, colloids, and suspensions—focusing on their physical properties and appropriate clinical applications. True solutions, such as normal saline and dextrose 5% in water (D5W), are stable, clear, and safe for intravenous use. Colloids, like albumin and lipid emulsions, contain larger particles that remain suspended and are used for plasma volume expansion in critical care. Suspensions, such as antacids or antibiotic syrups, contain visible particles that require shaking and are administered orally, not intravenously.

The chapter then introduces key units of concentration used in healthcare, including percent weight/volume (% w/v), percent volume/volume (% v/v), molarity (M), milliequivalents per liter (mEq/L), normality (N), and parts per million (ppm). Understanding these units is crucial for calculating dosages, preparing IV fluids, and interpreting lab results. The section on preparation techniques provides step-by-step instructions for making common clinical solutions like 0.9% sodium chloride and D5W, with an emphasis on sterile procedures, accurate measurements, and proper labeling.

Finally, the chapter explores clinical applications, including fluid resuscitation, medication dilution, and compatibility considerations. Nurses are reminded of the importance of checking IV compatibility, monitoring for adverse reactions, and tailoring fluid therapy based on patient needs. By mastering these principles, nursing professionals can ensure the safe and effective use of solutions in various healthcare settings.

EXERCISES

Multiple Choice Questions

1. Which of the following is considered a true solution suitable for intravenous use?
   A) 0.9% NaCl
   B) Albumin 25%
   C) Barium sulfate suspension
   D) Magnesium hydroxide
2. What unit is most appropriate for expressing electrolyte concentration in IV fluids?
   A) % v/v
   B) mEq/L
   C) ppm
   D) Molarity
3. Which of the following correctly describes a colloid solution?
   A) Completely transparent and filterable
   B) Contains particles larger than 1000 nm
   C) Shows the Tyndall effect and remains suspended
   D) Suitable only for oral administration
4. What is the normal concentration of sodium chloride in physiological saline?
   A) 5%
   B) 1%
   C) 0.45%
   D) 0.9%

True or False

1. Suspensions must be shaken well before administration to ensure even distribution.
2. Normality (N) accounts for the valence of the solute and is used in redox and acid-base calculations.
3. Colloid particles pass freely through semipermeable membranes.
4. Glucose solutions like D5W can serve as both a fluid and a source of energy.
5. It is acceptable to mix any two IV drugs as long as they are diluted.

Fill in the Blanks

1. The unit used to describe very low concentrations such as toxins in blood or water is ___________.
2. A nurse preparing 1 liter of 5% Dextrose (D5W) needs to dissolve ___________ grams of glucose.
3. Colloids are often used in critical care for ___________ expansion.
4. The formula C1V1=C2V2C1V1 = C2V2C1V1=C2V2 is used to perform ___________ calculations.
5. The concentration unit % w/v expresses grams of solute per ___________ mL of solution.

Deep Thinking / Critical Reasoning Questions

1. Why is it critical to understand the difference between isotonic, hypotonic, and hypertonic solutions in IV therapy? Give examples of when each might be used.
2. Explain how an error in calculating the concentration of a potassium chloride IV solution could lead to serious patient harm.
3. Discuss the advantages and potential risks of using colloid solutions instead of crystalloids for fluid resuscitation.
4. How does the concept of compatibility influence decisions about mixing medications or fluids in one IV line?

Case Studies

Case 1: IV Drug Preparation
A physician orders 20 mEq of potassium chloride (KCl) to be administered in 100 mL of 0.9% NaCl over 1 hour. The available KCl stock is labeled 2 mEq/mL.
– What steps should the nurse follow to prepare and safely administer this solution?
– What patient parameters should be monitored during the infusion?

Case 2: Fluid Selection
A 70-year-old patient with hyponatremia and signs of cerebral edema is admitted to the emergency department. The physician prescribes 3% NaCl.
– What are the chemical and clinical considerations for preparing and

administering this hypertonic solution?
– Why might a nurse prefer this over isotonic saline in this scenario?

Case 3: Pediatric Medication
A child requires a reconstituted antibiotic suspension for home use. The parent is unfamiliar with how to administer it.
– How should the nurse explain proper storage, shaking technique, and dosing to ensure safety and effectiveness?

Final Thoughts

As you reach the end of this book, one undeniable truth stands out: chemistry is far more than a foundational science; it is a vital clinical language woven into every aspect of nursing practice. Often, nurses use this language intuitively, without fully realizing that every IV fluid administered, every laboratory value analyzed, and every medication prepared involves fundamental chemical processes at work. Whether correcting electrolyte imbalances, interpreting arterial blood gases, preparing intravenous solutions, or responding to changes in a patient's consciousness or cardiac rhythm, you are actively applying chemistry to provide safe and effective care.

Throughout this text, you have been introduced to essential chemical principles that form the backbone of clinical nursing. You have learned how water moves between body compartments through osmosis, how solutions are prepared and measured with precision, the catalytic functions enzymes perform in metabolism, and the critical roles that electrolytes, acids, bases, and buffers play in maintaining physiological balance. These concepts were never meant to remain abstract or confined to theory; rather, they are practical tools designed to enhance your clinical reasoning and decision-making.

In clinical settings, this knowledge enables you to assess patients holistically, integrating chemical data with clinical signs and symptoms to develop effective intervention plans. You are encouraged to think critically—question unexpected lab results, seek correlations between biochemical values and patient presentations, and recognize subtle changes that may signal worsening conditions. This critical thinking is especially important when caring for vulnerable groups such as children, older adults, or critically ill patients, where small deviations can have significant consequences.

It is important to remember that mastery does not mean memorizing every chemical formula or performing complex calculations manually. Instead, it means cultivating the ability to interpret, apply, and communicate chemical information effectively within the clinical context. This empowers you to deliver care that is not only compassionate but scientifically sound and evidence-based.

Moreover, a solid grasp of chemistry allows you to bridge multiple disciplines — biology, pharmacology, pathophysiology, and clinical skills — creating a cohesive understanding that enhances your practice. In today's healthcare landscape, increasingly shaped by precision medicine, biotechnology, and advanced diagnostics, your ability to interpret chemical principles at the bedside becomes indispensable.

By engaging deeply with the material presented in this course, you have taken a critical step forward in your professional development. You are now better equipped to be not only a compassionate caregiver but also a clinically informed nursing professional with a strong scientific foundation. This knowledge strengthens your confidence and competence, enabling you to understand not just what actions to take, but why those actions matter — a vital distinction that elevates nursing from task-oriented care to a profession grounded in thoughtful, informed practice.

As you continue your nursing journey, let this integrated understanding of chemistry serve as a cornerstone for lifelong learning and professional excellence. Embrace the role of science in your daily work, and remember that your unique ability to translate complex chemical principles into compassionate, patient-centered care makes you an essential contributor to health and healing.

Appendices and Learning Tools

A. Glossary of Key Terms

Acid: A substance that increases the hydrogen ion (H^+) concentration in a solution.
Base: A substance that decreases the hydrogen ion concentration, often by releasing OH^- ions.
Buffer: A solution that resists changes in pH when acids or bases are added.
Carbohydrate: Organic compound composed of carbon, hydrogen, and oxygen; used primarily for energy.
Catalyst: A substance that speeds up a chemical reaction without being consumed.
Colloid: A mixture with particles larger than a true solution but that do not settle out.
Concentration: The amount of solute in a given volume of solution, often expressed in % w/v, molarity, or mEq/L.
Covalent Bond: A chemical bond formed when two atoms share one or more pairs of electrons.
Diffusion: Movement of molecules from an area of high concentration to low concentration.
Dilution: Reducing the concentration of a solute in solution by adding more solvent.
Electrolyte: A substance that dissociates into ions in solution and conducts electricity.
Enzyme: A biological catalyst that speeds up chemical reactions in the body.
Glucose: A simple sugar that is an important energy source in living organisms.
Hemodialysis: A medical procedure that filters blood using a machine to remove waste and excess fluids.
Ion: An atom or molecule with a net electric charge due to the loss or gain of one or more electrons.
Isotonic Solution: A solution with the same solute concentration as

body fluids, causing no net water movement.
Lipid: A group of hydrophobic biomolecules including fats, oils, and steroids; used for energy storage and insulation.
Molarity (M): Moles of solute per liter of solution; a measure of concentration.
Normality (N): Equivalents of solute per liter of solution; considers the reactivity of the solute.
Osmosis: The diffusion of water across a semipermeable membrane.
Plasma: The liquid portion of blood that carries cells and dissolved substances.
Protein: Large biomolecules made of amino acids; essential for structure and function in cells.
Solute: The substance that is dissolved in a solution.
Solution: A homogeneous mixture where solutes are completely dissolved in a solvent.
Solvent: The liquid in which the solute is dissolved, often water.
Suspension: A heterogeneous mixture where particles are dispersed but will eventually settle out.
Tonicity: The ability of a surrounding solution to cause a cell to gain or lose water.
Valence: The combining capacity of an element; determines how it bonds chemically.
pH: A scale measuring the hydrogen ion concentration of a solution; lower pH = more acidic.

B. Conversion Tables

The following conversion tables include commonly used units in healthcare and chemistry settings:

1. Mass

- 1 kilogram (kg) = 1000 grams (g)
- 1 gram (g) = 1000 milligrams (mg)
- 1 milligram (mg) = 1000 micrograms (μg)
- 1 microgram (μg) = 1000 nanograms (ng)

2. Volume

- 1 liter (L) = 1000 milliliters (mL)
- 1 milliliter (mL) = 1 cubic centimeter (cc)
- 1 teaspoon (tsp) = 5 mL
- 1 tablespoon (tbsp) = 15 mL
- 1 fluid ounce (oz) = 30 mL

3. Length

- 1 meter (m) = 100 centimeters (cm)
- 1 centimeter (cm) = 10 millimeters (mm)
- 1 inch = 2.54 centimeters

4. Temperature

- °C to °F: $(°C \times 9/5) + 32$
- °F to °C: $(°F - 32) \times 5/9$

5. Pressure

- 1 atmosphere (atm) = 760 mmHg
- 1 mmHg = 1 torr
- 1 atm = 101.3 kPa

6. Concentration

- 1% (w/v) = 1 gram of solute in 100 mL of solution
- 1 ppm = 1 mg/L in water
- 1 molarity (M) = 1 mole of solute per liter of solution

C. The Chapters' Answers

1 – Chemistry in Human Health and Clinical Care

Multiple Choice Questions (MCQs):

1. B 2. B 3. C 4. C 5. C

Short Answer Questions:

1. Maintains blood pH through buffering acids/bases.
2. Catalyze reactions by lowering activation energy.
3. Discovery, Preclinical, Clinical Trials (Phases I–III), Approval, Phase IV.
4. Developed drug techniques; introduced many medicines.
5. Supports dosage calculation, lab interpretation, and safe care.

Case-Based Problem: • Hypokalemia impairs nerve/muscle signals → arrhythmias. • Treat with potassium supplements and monitor levels.

True or False:

1. T 2. T 3. F 4. T 5. T

Fill in the Blanks:

1. composition, structure, properties, changes
2. oxygen, carbon, hydrogen, nitrogen, calcium, phosphorus
3. nerve impulses, muscle
4. Ebers 5. opium 6. Phase I 7. wounds
5. Water 9. catalysts 10. sodium, potassium

Bonus Application Question: 40 mEq ÷ 2 mEq/mL = 20 mL

Chapter 2: Measurements and Biomedical Applications

Review Questions

1. Standard units ensure consistency, accuracy, and safety in biomedical data and clinical communication.
2. Density affects fluid concentration and helps assess hydration and diagnose disorders.
3. Higher temperature increases metabolic activity by accelerating chemical reactions.
4. Pressure moves blood through vessels and air through the lungs.
5. Acid–base imbalances disrupt cellular function. Example: respiratory acidosis from CO_2 retention.
6. Accurate temperature helps detect fever or hypothermia and guide interventions.
7. Understanding metabolism supports nutrition planning, healing, and chronic disease care.

Calculations

1. 5.3 kg   2. 37°C   3. ~2.5× $[H^+]$ 4. 320 kcal

Fill in the Blanks

1. Pascal   2. 7.35–7.45   3. Kilocalories   4. °F = (°C × 9/5) + 32

Multiple Choice

1. C   2. B   3. C

True/False

1. F   2. T   3. T   4. F

Nutrition Calculations

1. 435 kcal   2. 480 kcal   3. 140 kcal   4. 10,460 kJ   5. 245 kcal

Clinical

1. 308 kcal (slightly over) 2. 135 g 3. 55.6 g fat

Advanced

1. 275 g C, 125 g P, 44.4 g F 2. 16.7 g fat

Chapter 3: Chemical Bonds and Matter in Healthcare

Multiple Choice

1. c **2.** b **3.** c **4.** b **5.** c **6.** c **7.** c **8.** b
9. d

True or False

1. False **2.** True **3.** False **4.** True **5.** False
6. False **7.** True **8.** True **9.** False

Fill in the Blanks

1. atomic number **2.** mass number **3.** Covalent **4.** like dissolves like
5. electrolytes **6.** Alpha **7.** gamma camera **8.** Iodine-131
9. linear accelerator

Short Answer Questions

1. Element: one type of atom. Compound: two or more bonded elements.
2. Isotopes: same protons, different neutrons. Ex: I-131 for thyroid treatment.
3. Ionic compounds dissolve due to ion–dipole interactions with water.
4. Polar molecules dissolve in water → improves drug absorption & distribution.

Critical Thinking

1. K^+ imbalance affects nerve signals → arrhythmia, cramps, weakness.
2. Plasma used in cauterization, sterilization, wound healing technologies.
3. Monitor solubility, use emulsions or non-aqueous carriers, IV if needed.
4. Benefit: stable binding. Drawback: reduced release/metabolism.
5. Different radiation types have different safety profiles and uses.
6. Pre: hydration, no pregnancy. Post: limit contact, flush fluids.
7. Imaging uses low-dose for detection; therapy uses high-dose to kill cells.

Chapter 4: Organic Molecules and Body Functions

Multiple Choice:

1. C   2. A   3. B   4. C   5. C

Short Answer Questions:

1. A functional group is a specific group of atoms within a molecule responsible for its characteristic chemical reactions. It defines the molecule's reactivity, polarity, and biological behavior.
2. An aldehyde has a carbonyl group at the end of the carbon chain (–CHO), while a ketone has the carbonyl group within the chain (C=O).
3. – Monohydric: Ethanol – used as a disinfectant
   – Dihydric: Ethylene glycol – used as solvent/antifreeze
   – Trihydric: Glycerol – used in IV fluids and skin products
4. Ketone bodies form during fat metabolism, especially during fasting or uncontrolled diabetes, and are elevated in diabetic ketoacidosis
5. Carboxylic acids act as weak acids, helping to buffer and maintain the pH of blood through proton donation.

Clinical Application:

1. The fruity breath odor is caused by acetone, a ketone body elevated in diabetic ketoacidosis.
2. Amino compounds are essential in neurotransmitter synthesis; e.g., dopamine from tyrosine regulates mood and movement.
3. Aspirin contains an ester group, which affects its solubility, absorption, and anti-inflammatory action.
4. The hydroxyl group (–OH) in ethanol is responsible for its antiseptic properties by disrupting microbial membranes.
5. The carboxylic acid group increases solubility in water and influences the ionization and absorption of drugs in the body.

Complete the Sentence:

1. hydroxyl (–OH) 2. within the carbon chain 3. glycerol 4. acids

5. amino and carboxyl 6. formaldehyde 7. acetone 8. carboxylic acid and alcohol

True or False:

1. True 2. False 3. True 4. True 5. False 6. True 7. False 8. True 9. True 10. True

Chapter 5: Introduction to Biochemistry for Healthcare

Multiple Choice Questions (MCQs):

1-C, 2-B, 3-C, 4-D, 5-C

True or False:

1-F, 2-F, 3-T, 4-F, 5-F

Short Answer Questions:

1. Carbohydrates: energy, Proteins: structure, Lipids: membranes/energy, Nucleic acids: genetics
2. Essential fatty acids are required fats the body can't produce; vital for cell function and hormone synthesis
3. ATP stores and provides energy for cellular processes
4. Catabolism: breakdown & releases energy; Anabolism: builds & requires energy
5. Enzymes speed up chemical reactions in the body

Matching:

A-3, B-2, C-1, D-4, E-5

Fill in the Blanks:

1-Amino acids, 2-Glycolysis, 3-Thymine & Guanine, 4-Cholesterol, 5-Polysaccharide

Case Study 1 (Fatimah):

- Impaired pathway: Glucose metabolism
- Ketones: Body using fat for energy; possible ketoacidosis
- Protein metabolism: Increased breakdown for energy
- Nursing: Monitor glucose/ketones, give insulin, hydrate, educate on diet/management.

Case Study 2 (Ahmad):

- Deficiency: Protein (kwashiorkor)
- Albumin role: Maintains fluid balance & transports substances
- Danger: Edema, delayed healing, weak immunity
- Nursing: High-protein diet, monitor labs, refer to nutritionist, educate family

Case Study 3 (Samer):

- Most concerning: High LDL (180 mg/dL)

- Effect: LDL causes plaque buildup in arteries (atherosclerosis)
- Advice: Low-fat diet, more fiber, quit smoking, exercise
- Nursing: Monitor lipids, support lifestyle changes, medication education, assess CVD risk

Chapter 6: Bioactive Substances in the Body

Multiple Choice (MCQs)

1. C 2. B 3. C 4. C

True or False

1. True 2. False 3. False 4. True

Fill in the Blanks

1. Insulin 2. Sodium 3. Megaloblastic 4. Hypertension and hypokalemia

Short Answer & Critical Thinking

1. **Potassium levels should be monitored** because loop diuretics (e.g., furosemide) increase urinary excretion of potassium, which may lead to **hypokalemia**, causing arrhythmias and muscle weakness. Monitoring prevents complications and guides supplementation.
2. **Vitamin K is given to newborns** because they are born with low vitamin K levels and sterile intestines (no gut flora to produce it). This prevents **hemorrhagic disease of the newborn**, a potentially serious bleeding disorder.

Clinical Case Studies

Case 1

- **Likely condition**: Hypokalemia
- **Nursing interventions**:
    - Notify the physician

  - Administer prescribed potassium replacement (oral or IV)
  - Monitor cardiac rhythm and repeat potassium levels
  - Educate on dietary potassium if applicable

Case 2

- **Most likely diagnosis**: Hypothyroidism
- **Nursing education**:
  - Explain the importance of **daily thyroid hormone replacement (levothyroxine)**
  - Stress the need for **regular blood tests** to monitor TSH and T4
  - Educate about managing **weight gain, cold intolerance, and fatigue** through lifestyle changes
  - Advise taking the medication **on an empty stomach**, preferably in the morning

Chapter 7 : Water, pH, and Fluid Balance

Multiple Choice:

1. A 2. D 3. B 4. B 5. C

True or False:

1. True 2. False 3. False 4. False 5. True

Fill in the Blanks:

1. 7.35 to 7.45 2. Hydrostatic 3. ADH 4. Osmosis 5. Oncotic

Critical Thinking Questions:

1. **Buffer systems** help maintain acid–base balance by immediately neutralizing excess hydrogen (H^+) or hydroxide (OH^-) ions, stabilizing pH levels. The bicarbonate buffer is the most important in blood, preventing drastic pH shifts that could impair enzyme function and cellular metabolism.

2. **Dehydration** reduces extracellular fluid volume, increasing solute concentration and causing water to leave cells. This disrupts cell function, impairs circulation, and can lead to electrolyte imbalances, altered mental status, and organ dysfunction.
3. A nurse should assess fluid status by monitoring vital signs (e.g., hypotension, tachycardia), urine output, mucous membranes, skin turgor, weight changes, and lab values such as sodium, hematocrit, and BUN. Prompt replacement of fluids and electrolytes is essential.

Clinical Case Scenarios:

1. **Case 1** – Likely condition: **Dehydration with hypernatremia**.
   Nursing interventions: Initiate hypotonic IV fluids (e.g., 0.45% NaCl), monitor sodium levels, vital signs, mental status, and urine output. Encourage oral fluids if possible.
2. **Case 2** – Likely condition: **Fluid overload and hyponatremia**.
   Next steps: Slow or stop IV fluids, raise the head of the bed, monitor respiratory status and lung sounds, check electrolyte levels, and notify the physician. Diuretics may be ordered.
3. **Case 3** – Likely condition: **Respiratory acidosis** due to CO_2 retention (common in COPD).
   Care plan adjustment: Support ventilation (e.g., oxygen with caution in COPD), assess ABGs regularly, elevate head of bed, and encourage effective breathing techniques. Avoid over-oxygenation that can suppress respiratory drive.

Chapter 8: Solutions in Clinical Practice

Multiple Choice Questions (MCQs):

1. A 2. B 3. C 4. D

True or False:

1. True 2. True 3. False 4. True 5. False

Fill in the Blanks:

1. ppm , ppb 2. 50 3. plasma volume 4. dilution
   5. 100

Deep Thinking / Critical Reasoning Questions:

1. Understanding tonicities is crucial in managing fluid shifts. Isotonic (e.g., 0.9% NaCl) maintains balance; hypotonic (e.g., 0.45% NaCl) hydrates cells; hypertonic (e.g., 3% NaCl) reduces edema.
2. Potassium errors can cause lethal hyperkalemia or arrhythmias. Precise calculations and infusion control are vital.
3. Colloids offer longer intravascular volume support but may cause allergic reactions, coagulation issues, or renal strain.
4. Incompatible IV mixtures may precipitate or inactivate drugs. Always confirm compatibility to ensure safety.

Case Studies:

Case 1: KCl in 0.9% NaCl
– Use 10 mL of KCl (2 mEq/mL) + 90 mL of 0.9% NaCl
– Monitor: ECG, serum potassium, renal function, infusion site

Case 2: 3% NaCl for cerebral edema
– Hypertonic solution used to treat hyponatremia and cerebral swelling
– Preferred over isotonic saline because it draws fluid out of brain tissue

Case 3: Antibiotic suspension (pediatric)
– Store in the refrigerator, shake well before use, use a calibrated measuring device, complete the full course, discard after 7–10 days.

D. Safety Guidelines

The following safety practices should be followed in any clinical or laboratory environment:

1. General Laboratory Safety

- Always wear personal protective equipment (PPE): gloves, lab coat, and safety goggles.
- Tie back long hair and avoid loose clothing.
- Do not eat, drink, or apply cosmetics in laboratory areas.
- Know the location of emergency equipment: eyewash stations, fire extinguishers, and first aid kits.

2. Chemical Handling

- Read all labels and safety data sheets (SDS) before using chemicals.
- Label all containers clearly with contents, concentration, and date of preparation.
- Never mix chemicals unless instructed to do so.
- Handle volatile substances in a fume hood.
- Store chemicals according to compatibility (e.g., acids separate from bases).

3. Biological and Clinical Safety

- Use standard precautions when handling blood or body fluids.
- Dispose of sharps in designated puncture-proof containers.
- Decontaminate work surfaces after use.
- Report spills or exposures immediately to the supervisor.

4. Waste Management

- Segregate chemical, biological, and sharps waste.
- Use color-coded containers for different types of waste.
- Follow institutional protocols for hazardous waste pickup and documentation.

E. Acknowledgment

The author gratefully acknowledges the invaluable support and insights provided by colleagues, students, and academic peers who contributed constructive feedback throughout the preparation of this textbook. Special thanks are extended to **Dr. Maher El Bayome** and the Faculty of Nursing and Health Sciences at **Israa University – Gaza** for their unwavering encouragement and academic collaboration.

The author also expresses sincere appreciation to international partners and collaborators involved in educational and capacity-building programs, whose contributions have significantly enriched the content and clinical applicability of this work.

Recognition is given to the contributions of modern AI tools such as **ChatGPT by OpenAI** and **Canva**, which were instrumental in refining the content, improving the language, and enhancing the visual and structural layout of the manuscript, particularly during the final editing phase.

This book is dedicated to students and healthcare professionals who tirelessly pursue the advancement of human health through knowledge, compassion, and scientific excellence.

F. References

Books

1. Timberlake, K. C. (2024). *Chemistry: An introduction to general, organic, and biological chemistry* (14th ed.). Pearson.
2. Tro, N. J. (2021). *Introductory chemistry* (4th ed.). Pearson. *(Publisher added)*

Educational Websites

1. ChemCollective. (n.d.). *Interactive virtual labs and simulations*. Retrieved June 26, 2025, from http://chemcollective.org
2. Khan Academy. (n.d.). *Chemistry and biology lessons*. Retrieved June 26, 2025, from https://www.khanacademy.org
3. MedlinePlus. (n.d.). *Health information*. U.S. National Library of Medicine. Retrieved June 26, 2025, from https://medlineplus.gov
4. PubChem. (n.d.). *Chemical database*. National Center for Biotechnology Information. Retrieved June 26, 2025, from https://pubchem.ncbi.nlm.nih.gov
5. ChemSpider. (n.d.). *Free chemical structure database*. Royal Society of Chemistry. Retrieved June 26, 2025, from http://www.chemspider.com

Professional Organizations

1. American Chemical Society. (2025). *ACS: Advancing chemistry education and research*. Retrieved June 26, 2025, from https://www.acs.org
2. American Nurses Association. (2025). *ANA: Professional nursing resource hub*. Retrieved June 26, 2025, from https://www.nursingworld.org
3. International Council of Nurses. (2025). *ICN: Global nursing standards*. Retrieved June 26, 2025, from https://www.icn.ch

Journal Articles & Reports

1. Borrelli, E. P., Lucaci, J. D., Wilson, N. S., Taneja, A., Weiss, M., & Beer, I. (2025). Evaluating the impact of smart infusion pump interoperability on reducing medication administration errors: A systematic literature review. *Medical Devices: Evidence and Research, 18*, 247–260. https://doi.org/10.2147/MDER.S522534 *(Corrected journal title, volume formatting, and removed PMID/PMCID)*
2. Białecka-Dębek, A., Madej, D., & Łojek, E. (2025). Water intake, hydration status and cognitive functions in older adults – a pilot study. *European Journal of Nutrition*. Advance online publication. https://doi.org/10.1007/s00394-025-03690-1
3. Ciortea, D.-A., et al. (2024). Impact of hyponatremia and ADH secretion in MIS-C and COVID-19: An integrative

approach. *Current Issues in Molecular Biology, 46*(11), 11749–11771. https://doi.org/10.3390/cimb46110698

4. Rich, D. S., Fricker, M. P., Jr., Cohen, M. R., & Levine, S. R. (2013). Guidelines for the safe preparation of sterile compounds: Results of the ISMP sterile preparation compounding safety summit of October 2011. *Hospital Pharmacy, 48*(4), 282–294. https://doi.org/10.1310/hpj4804-282

(Added Jr. formatting and corrected journal title)

5. Sedlacek, M., Schoolwerth, A. C., & Remillard, B. D. (2006). Electrolyte disturbances in the intensive care unit. *Seminars in Dialysis, 19*(6), 496–501. https://doi.org/10.1111/j.1525-139X.2006.00212.x
6. Zhang, L., & Xia, Y. (2023). Bioorthogonal chemistry for prodrug activation in vivo. *Chemical Society Reviews, 52*(5), 2100–2120. https://doi.org/10.1039/D2CS00889K
7. Paoli, A., et al. (2023). Emerging pathophysiological roles of ketone bodies. *Physiology, 38*(6), 363–376. https://doi.org/10.1152/physiol.00031.2023
8. Wang, J., et al. (2024). Stable porous organic cage nanocapsules for pH responsive paclitaxel delivery. *ACS Applied Bio Materials, 7*(8), 1234–1245. https://doi.org/10.1021/acsabm.4c01123
9. Zulet, M. Á., et al. (2023). Water intake, hydration status and 2 year changes in cognitive performance among older adults with metabolic syndrome. *BMC Medicine, 21*, 177. https://doi.org/10.1186/s12916-023-02771-4

Periodic Table of the Elements

1	2	3	4	5	6	7	8	9	10	11	12	13	14	15	16	17	18
1 H Hydrogen 1.01																	2 He Helium 4.00
3 Li Lithium 6.94	4 Be Beryllium 9.01											5 B Boron 10.81	6 C Carbon 12.01	7 N Nitrogen 14.01	8 O Oxygen 16.00	9 F Fluorine 19.00	10 Ne Neon 20.18
11 Na Sodium 22.99	12 Mg Magnesium 24.31											13 Al Aluminum 26.98	14 Si Silicon 28.09	15 P Phosphorus 30.97	16 S Sulfur 32.06	17 Cl Chlorine 35.45	18 Ar Argon 39.95
19 K Potassium 39.10	20 Ca Calcium 40.08	21 Sc Scandium 44.96	22 Ti Titanium 47.88	23 V Vanadium 50.94	24 Cr Chromium 51.99	25 Mn Manganese 54.94	26 Fe Iron 55.85	27 Co Cobalt 58.93	28 Ni Nickel 58.69	29 Cu Copper 63.55	30 Zn Zinc 65.38	31 Ga Gallium 69.72	32 Ge Germanium 72.63	33 As Arsenic 74.92	34 Se Selenium 78.97	35 Br Bromine 79.90	36 Kr Krypton 83.80
37 Rb Rubidium 85.47	38 Sr Strontium 87.62	39 Y Yttrium 88.91	40 Zr Zirconium 91.22	41 Nb Niobium 92.91	42 Mo Molybdenum 95.95	43 Tc Technetium 98.91	44 Ru Ruthenium 101.07	45 Rh Rhodium 102.91	46 Pd Palladium 106.42	47 Ag Silver 107.87	48 Cd Cadmium 112.41	49 In Indium 114.82	50 Sn Tin 118.71	51 Sb Antimony 121.76	52 Te Tellurium 127.6	53 I Iodine 126.90	54 Xe Xenon 131.29
55 Cs Cesium 132.91	56 Ba Barium 137.33	57-71 Lanthanides	72 Hf Hafnium 178.49	73 Ta Tantalum 180.95	74 W Tungsten 183.85	75 Re Rhenium 186.21	76 Os Osmium 190.23	77 Ir Iridium 192.22	78 Pt Platinum 195.08	79 Au Gold 196.97	80 Hg Mercury 200.59	81 Tl Thallium 204.38	82 Pb Lead 207.20	83 Bi Bismuth 208.98	84 Po Polonium [208.98]	85 At Astatine 209.98	86 Rn Radon 222.02
87 Fr Francium 223.02	88 Ra Radium 226.03	89-103 Actinides	104 Rf Rutherfordium [261]	105 Db Dubnium [262]	106 Sg Seaborgium [266]	107 Bh Bohrium [264]	108 Hs Hassium [269]	109 Mt Meitnerium [278]	110 Ds Darmstadtium [281]	111 Rg Roentgenium [280]	112 Cn Copernicium [285]	113 Nh Nihonium [286]	114 Fl Flerovium [289]	115 Mc Moscovium [289]	116 Lv Livermorium [293]	117 Ts Tennessine [294]	118 Og Oganesson [294]

57 La Lanthanum 138.91	58 Ce Cerium 140.12	59 Pr Praseodymium 140.91	60 Nd Neodymium 144.24	61 Pm Promethium 144.91	62 Sm Samarium 150.36	63 Eu Europium 151.96	64 Gd Gadolinium 157.25	65 Tb Terbium 158.93	66 Dy Dysprosium 162.50	67 Ho Holmium 164.93	68 Er Erbium 167.26	69 Tm Thulium 168.93	70 Yb Ytterbium 173.06	71 Lu Lutetium 174.97
89 Ac Actinium 227.03	90 Th Thorium 232.04	91 Pa Protactinium 231.04	92 U Uranium 238.03	93 Np Neptunium 237.05	94 Pu Plutonium 244.06	95 Am Americium 243.06	96 Cm Curium 247.07	97 Bk Berkelium 247.07	98 Cf Californium 251.08	99 Es Einsteinium [254]	100 Fm Fermium 257.10	101 Md Mendelevium 258.10	102 No Nobelium 259.10	103 Lr Lawrencium [262]

Alkali Metal | Alkaline Earth | Transition Metal | Basic Metal | Metalloid | Nonmetal | Halogen | Noble Gas | Lanthanide | Actinide

FSC
www.fsc.org
MIX
Papier | Fördert gute Waldnutzung
FSC® C083411

Kolibri 360 GmbH
Ferdinand-Jühlke-Straße 7
99095 Erfurt, Deutschland
produktsicherheit@kolibri360.de

Druck:
CPI Druckdienstleistungen GmbH
im Auftrag der
Kolibri 360 GmbH
Ferdinand-Jühlke-Str. 7
99095 Erfurt